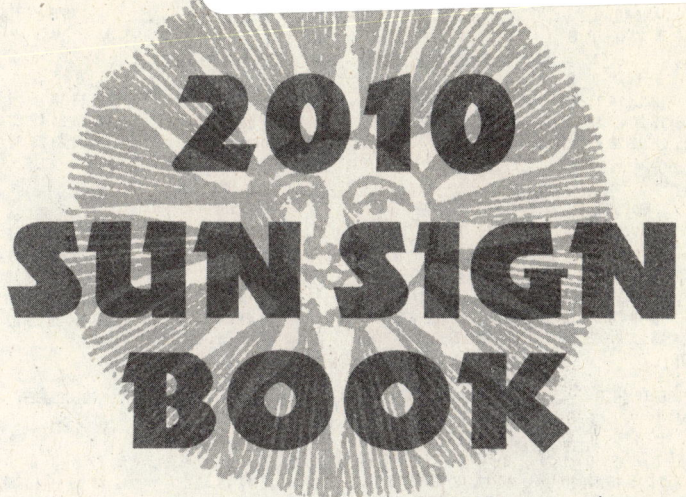

Forecasts by
Kris Brandt Riske

Book Editing: Sharon Leah
Cover Design: Kevin R. Brown

Copyright 2009 Llewellyn Publications
ISBN: 978-0-7387-0689-4
A Division of Llewellyn Worldwide, Ltd.
Llewellyn is a registered trademark of Llewellyn Worldwide, Ltd.
2143 Wooddale Drive, Woodbury, MN 55125
Printed in the USA

2010

JANUARY
S	M	T	W	T	F	S
					1	2
3	4	5	6	7	8	9
10	11	12	13	14	15	16
17	18	19	20	21	22	23
24	25	26	27	28	29	30
31						

FEBRUARY
S	M	T	W	T	F	S
	1	2	3	4	5	6
7	8	9	10	11	12	13
14	15	16	17	18	19	20
21	22	23	24	25	26	27
28						

MARCH
S	M	T	W	T	F	S
	1	2	3	4	5	6
7	8	9	10	11	12	13
14	15	16	17	18	19	20
21	22	23	24	25	26	27
28	29	30	31			

APRIL
S	M	T	W	T	F	S
				1	2	3
4	5	6	7	8	9	10
11	12	13	14	15	16	17
18	19	20	21	22	23	24
25	26	27	28	29	30	

MAY
S	M	T	W	T	F	S
						1
2	3	4	5	6	7	8
9	10	11	12	13	14	15
16	17	18	19	20	21	22
23	24	25	26	27	28	29
30	31					

JUNE
S	M	T	W	T	F	S
		1	2	3	4	5
6	7	8	9	10	11	12
13	14	15	16	17	18	19
20	21	22	23	24	25	26
27	28	29	30			

JULY
S	M	T	W	T	F	S
				1	2	3
4	5	6	7	8	9	10
11	12	13	14	15	16	17
18	19	20	21	22	23	24
25	26	27	28	29	30	31

AUGUST
S	M	T	W	T	F	S
1	2	3	4	5	6	7
8	9	10	11	12	13	14
15	16	17	18	19	20	21
22	23	24	25	26	27	28
29	30	31				

SEPTEMBER
S	M	T	W	T	F	S
			1	2	3	4
5	6	7	8	9	10	11
12	13	14	15	16	17	18
19	20	21	22	23	24	25
26	27	28	29	30		

OCTOBER
S	M	T	W	T	F	S
					1	2
3	4	5	6	7	8	9
10	11	12	13	14	15	16
17	18	19	20	21	22	23
24	25	26	27	28	29	30
31						

NOVEMBER
S	M	T	W	T	F	S
	1	2	3	4	5	6
7	8	9	10	11	12	13
14	15	16	17	18	19	20
21	22	23	24	25	26	27
28	29	30				

DECEMBER
S	M	T	W	T	F	S
			1	2	3	4
5	6	7	8	9	10	11
12	13	14	15	16	17	18
19	20	21	22	23	24	25
26	27	28	29	30	31	

2011

JANUARY
S	M	T	W	T	F	S
						1
2	3	4	5	6	7	8
9	10	11	12	13	14	15
16	17	18	19	20	21	22
23	24	25	26	27	28	29
30	31					

FEBRUARY
S	M	T	W	T	F	S
		1	2	3	4	5
6	7	8	9	10	11	12
13	14	15	16	17	18	19
20	21	22	23	24	25	26
27	28					

MARCH
S	M	T	W	T	F	S
		1	2	3	4	5
6	7	8	9	10	11	12
13	14	15	16	17	18	19
20	21	22	23	24	25	26
27	28	29	30	31		

APRIL
S	M	T	W	T	F	S
					1	2
3	4	5	6	7	8	9
10	11	12	13	14	15	16
17	18	19	20	21	22	23
24	25	26	27	28	29	30

MAY
S	M	T	W	T	F	S
1	2	3	4	5	6	7
8	9	10	11	12	13	14
15	16	17	18	19	20	21
22	23	24	25	26	27	28
29	30	31				

JUNE
S	M	T	W	T	F	S
			1	2	3	4
5	6	7	8	9	10	11
12	13	14	15	16	17	18
19	20	21	22	23	24	25
26	27	28	29	30		

JULY
S	M	T	W	T	F	S
					1	2
3	4	5	6	7	8	9
10	11	12	13	14	15	16
17	18	19	20	21	22	23
24	25	26	27	28	29	30
31						

AUGUST
S	M	T	W	T	F	S
	1	2	3	4	5	6
7	8	9	10	11	12	13
14	15	16	17	18	19	20
21	22	23	24	25	26	27
28	29	30	31			

SEPTEMBER
S	M	T	W	T	F	S
				1	2	3
4	5	6	7	8	9	10
11	12	13	14	15	16	17
18	19	20	21	22	23	24
25	26	27	28	29	30	

OCTOBER
S	M	T	W	T	F	S
						1
2	3	4	5	6	7	8
9	10	11	12	13	14	15
16	17	18	19	20	21	22
23	24	25	26	27	28	29
30	31					

NOVEMBER
S	M	T	W	T	F	S
		1	2	3	4	5
6	7	8	9	10	11	12
13	14	15	16	17	18	19
20	21	22	23	24	25	26
27	28	29	30			

DECEMBER
S	M	T	W	T	F	S
				1	2	3
4	5	6	7	8	9	10
11	12	13	14	15	16	17
18	19	20	21	22	23	24
25	26	27	28	29	30	31

Table of Contents

Meet Kris Brandt Riske 5
New Concepts for Signs of the Zodiac 6
Understanding the Basics of Astrology 8
Signs of the Zodiac 9
The Planets 10
Using this Book 11
2010 at a Glance 12
Ascendant Table 14
Astrological Glossary 16
Meanings of the Planets 22

2010 Sun Sign Book Forecasts

Aries 28
Taurus 51
Gemini 73
Cancer 96
Leo 119
Virgo 142
Libra 166
Scorpio 190
Sagittarius 214
Capricorn 238
Aquarius 262
Pisces 286

2010 Sun Sign Book Articles

The Mystery of Healing Gems
  by Shakti Navran 313

The Astrology of Social Networking Sites
  by Robin Ivy 322

Leaders Follow the Sun
  by Bruce Scofield 331

Home Makeovers on a Streamlined Budget
  by Alice DeVille 341

Sunspot Explosion in 2010
  by Kaye Shinker 347

Sun Sign's Orientation to Change
  by Lesley Francis 353

About the Contributors 360

Meet Kris Brandt Riske

Kris Brandt Riske is the executive director and a professional member of the American Federation of Astrologers (AFA), the oldest U.S. astrological organization, founded in 1938; and a member of National Council for Geocosmic Research (NCGR). She has a master's degree in journalism and is currently a student in the Penn State weather forecasting program.

Kris is the author of several books, including: *Llewellyn's Complete Book of Astrology: The Easy Way to Learn Astrology*; *Mapping Your Money*; *Mapping Your Future*; and coauthor of *Mapping Your Travels and Relocation*; and *Astrometeorology: Planetary Powers in Weather Forecasting*. She also writes for astrology publications, and does the annual weather forecast for *Llewellyn's Moon Sign Book*. In addition to astrometeorology, she specializes in predictive astrology. Kris is an avid NASCAR fan, although she'd rather be a driver than a spectator. She also enjoys gardening, reading, jazz, and her three cats.

New Concepts for Signs of the Zodiac

The signs of the zodiac represent characteristics and traits that indicate how energy operates within our lives. The signs tell the story of human evolution and development, and all are necessary to form the continuum of whole life experience. In fact, all twelve signs are represented within your astrological chart.

Although the traditional metaphors for the twelve signs (such as Aries, the Ram) are always functional, these alternative concepts for each of the twelve signs also describe the gradual unfolding of the human spirit.

Aries: The Initiator is the first sign of the zodiac and encompasses the primary concept of getting things started. This fiery ignition and bright beginning can prove to be the thrust necessary for new life, but the Initiator also can appear before a situation is ready for change and create disruption.

Taurus: The Maintainer sustains what Aries has begun and brings stability and focus into the picture, yet there also can be a tendency to try to maintain something in its current state without allowing for new growth.

Gemini: The Questioner seeks to determine whether alternatives are possible and offers diversity to the processes Taurus has brought into stability. Yet questioning can also lead to distraction, subsequently scattering energy and diffusing focus.

Cancer: The Nurturer provides the qualities necessary for growth and security, and encourages a deepening awareness of emotional needs. Yet this same nurturing can stifle individuation if it becomes too smothering.

Leo: The Loyalist directs and centralizes the experiences Cancer feeds. This quality is powerfully targeted toward self-awareness, but can be shortsighted. Hence, the Loyalist can hold steadfastly to viewpoints or feelings that inhibit new experiences.

Virgo: The Modifier analyzes the situations Leo brings to light and determines possibilities for change. Even though this change may be in the name of improvement, it can lead to dissatisfaction with the self if not directed in harmony with higher needs.

Libra: The Judge is constantly comparing everything to be sure that a certain level of rightness and perfection is presented. However, the Judge can also present possibilities that are harsh and seem to be cold or without feeling.

Scorpio: The Catalyst steps into the play of life to provide the quality of alchemical transformation. The Catalyst can stir the brew just enough to create a healing potion, or may get things going to such a powerful extent that they boil out of control.

Sagittarius: The Adventurer moves away from Scorpio's dimension to seek what lies beyond the horizon. The Adventurer continually looks for possibilities that answer the ultimate questions, but may forget the pathway back home.

Capricorn: The Pragmatist attempts to put everything into its rightful place and find ways to make life work out right. The Pragmatist can teach lessons of practicality and determination, but can become highly self-righteous when shortsighted.

Aquarius: The Reformer looks for ways to take what Capricorn has built and bring it up to date. Yet there is also a tendency to scrap the original in favor of a new plan that may not have the stable foundation necessary to operate effectively.

Pisces: The Visionary brings mysticism and imagination, and challenges the soul to move beyond the physical plane, into the realm of what might be. The Visionary can pierce the veil, returning enlightened to the physical world. The challenge is to avoid getting lost within the illusion of an alternate reality.

Understanding the Basics of Astrology

Astrology is an ancient and continually evolving system used to clarify your identity and your needs. An astrological chart—which is calculated using the date, time, and place of birth—contains many factors that symbolically represent the needs, expressions, and experiences that make up the whole person. A professional astrologer interprets this symbolic picture, offering you an accurate portrait of your personality.

The chart itself—the horoscope—is a portrait of an individual. Generally, a natal (or birth) horoscope is drawn on a circular wheel. The wheel is divided into twelve segments, called houses. Each of the twelve houses represents a different aspect of the individual, much like the facets of a brilliantly cut stone. The houses depict different environments, such as home, school, and work. The houses also represent roles and relationships: parents, friends, lovers, children, partners. In each environment, individuals show a different side of their personality. At home, you may represent yourself quite differently than you do on the job. Additionally, in each relationship you will project a different image of yourself. Your parents rarely see the side you show to intimate friends.

Symbols for the planets, the Sun, and the Moon are drawn inside the houses. Each planet represents a separate kind of energy. You experience and express that energy in specific ways. (For a complete list, refer to the table on the next page.) The way you use each of these energies is up to you. The planets in your chart do not make you do anything!

The twelve signs of the zodiac indicate characteristics and traits that further define your personality. Each sign can be expressed in positive and negative ways. (The basic meaning of each of the signs is explained in the corresponding sections ahead.) What's more, you have all twelve signs somewhere in your chart. Signs that are strongly emphasized by the planets have greater force. The Sun, Moon, and planets are placed on the chart according to their position at the time of birth. The qualities of a sign, combined with the energy of a planet, indicate how you might be most likely to use

Signs of the Zodiac

Aries	♈	The Initiator
Taurus	♉	The Maintainer
Gemini	♊	The Questioner
Cancer	♋	The Nurturer
Leo	♌	The Loyalist
Virgo	♍	The Modifier
Libra	♎	The Judge
Scorpio	♏	The Catalyst
Sagittarius	♐	The Adventurer
Capricorn	♑	The Pragmatist
Aquarius	♒	The Reformer
Pisces	♓	The Visionary

that energy and the best ways to develop that energy. The signs add color, emphasis, and dimension to the personality.

Signs are also placed at the cusps, or dividing lines, of each of the houses. The influence of the signs on the houses is much the same as their influence on the Sun, Moon, and planets. Each house is shaped by the sign on its cusp.

When you view a horoscope, you will notice that there appear to be four distinctive angles dividing the wheel of the chart. The line that divides the chart into a top and bottom half represents the horizon. In most cases, the left side of the horizon is called the Ascendant. The zodiac sign on the Ascendant is your rising sign. The Ascendant indicates the way others are likely to view you.

The Sun, Moon, or planet can be compared to an actor in a play. The sign shows how the energy works, like the role the actor plays in a drama. The house indicates where the energy operates, like the setting of a play. On a psychological level, the Sun represents who you think you are. The Ascendant describes who others think you are, and the Moon reflects your inner self.

The Planets

Sun	☉	The ego, self, willpower
Moon	☽	The subconscious self, habits
Mercury	☿	Communication, the intellect
Venus	♀	Emotional expression, love, appreciation, artistry
Mars	♂	Physical drive, assertiveness, anger
Jupiter	♃	Philosophy, ethics, generosity
Saturn	♄	Discipline, focus, responsibility
Uranus	♅	Individuality, rebelliousness
Neptune	♆	Imagination, sensitivity, compassion
Pluto	♇	Transformation, healing, regeneration

Astrologers also study the geometric relationships between the Sun, Moon, and planets. These geometric angles are called aspects. Aspects further define the strengths, weaknesses, and challenges within your physical, mental, emotional, and spiritual self. Sometimes, patterns also appear in an astrological chart. These patterns have meaning.

To understand cycles for any given point in time, astrologers study several factors. Many use transits, which refer to the movement and positions of the planets. When astrologers compare those positions to the birth horoscope, the transits indicate activity in particular areas of the chart. The *Sun Sign Book* uses transits.

As you can see, your Sun sign is just one of many factors that describe who you are—but it is a powerful one! As the symbol of the ego, the Sun in your chart reflects your drive to be noticed. Most people can easily relate to the concepts associated with their Sun sign, since it is tied to their sense of personal identity.

Using this Book

This book contains what is called "Sun sign astrology," that is, astrology based on the sign that your Sun was in at the time of your birth. The technique has its foundation in ancient Greek astrology, in which the Sun was one of five points in the chart that was used as a focal point for delineation.

The most effective way to use astrology, however, is through one-on-one work with a professional astrologer, who can integrate the eight or so other astrological bodies into the interpretation to provide you with guidance. There are factors related to the year and time of day you were born that are highly significant in the way you approach life and vital to making wise choices. In addition, there are ways of using astrology that aren't addressed here, such as compatibility between two specific individuals, discovering family patterns, or picking a day for a wedding or grand opening.

To best use the information in the monthly forecasts, you'll want to determine your Ascendant, or rising sign. If you don't know your Ascendant, the tables following this description will help you determine your rising sign. They are most accurate for those born in the continental United States. They're only an approximation, but they can be used as a good rule of thumb. Your exact Ascendant may vary from the tables according to your time and place of birth. Once you've approximated your ascending sign using the tables or determined your Ascendant by having your chart calculated, you'll know two significant factors in your chart. Read the monthly forecast sections for both your Sun and Ascendant to gain the most useful information. In addition, you can read the section about the sign your Moon is in. The Sun is the true, inner you; the Ascendant is your shell or appearance and the person you are becoming; the Moon is the person you were—or still are based on habits and memories.

I've also included information about the planets' retrogrades this year. Most people have heard of "Mercury retrograde." In fact, all the planets except the Sun and Moon appear to travel backward (retrograde) in their path periodically. This only appears to happen because we on the Earth are not seeing the other planets from the middle of the solar system. Rather, we are watching them from our own moving object. We are like a train that moves past cars on the freeway that are going at a slower speed. To us on the train, the cars look like they're going backward. Mercury turns retrograde about every four months for three weeks; Venus every eighteen months for six weeks; Mars every two years for two to three months. The rest of the planets each retrograde once a year for four to five months. During each retrograde, we have the opportunity to try something new, something we conceived of at the beginning of the planet's yearly cycle. The times when the planets change direction are significant, as are the beginning and midpoint (peak or culmination) of each cycle. These are noted in your forecast each month.

Your "Rewarding and Challenging Days" sections indicate times when you'll feel either more centered or more out of balance. The rewarding days are not the only times you can perform well, but the times you're likely to feel better integrated! During challenging days, take extra time to center yourself by meditating or using other techniques that help you feel more objective.

The Action Table found at the end of each sign's section offers general guidelines for the best time to take a particular action. Please note, however, that your whole chart will provide more accurate guidelines for the best time to do something. Therefore, use this table with a grain of salt, and never let it stop you from taking an action you feel compelled to take.

You can use this information for an objective awareness about the way the current cycles are affecting you. Realize that the power of astrology is even more useful when you have a complete chart and professional guidance.

2010 at a Glance

Four of the outer planets—Jupiter, Saturn, Uranus, and Pluto—are major players in 2010, just as they have been the past two years. Whenever two (or more) of these planets join forces they influence global conditions and also individual lives, depending upon their zodiacal placement.

Change and instability continue this year as these four planets align individually and collectively. Saturn, which entered Libra late last fall, briefly returns to Virgo to complete unfinished business in that sign. Jupiter and Uranus spend much of the year in

Pisces, visiting Aries for about three months. Pluto continues to inch forward in Capricorn, the sign it entered in 2008, and will occupy until 2024.

This year brings the final alignment of Saturn in Virgo and Uranus in Pisces. (The others were in November 2008, February 2009, and September 2009.) This push-pull of opposing forces is a challenging one to resolve. Change does not mesh well with the status quo, yet this is exactly what this planetary duo aims to achieve: a new and very different reality—restructuring on a global and personal scale. But there's more. Saturn and Uranus clash again this summer, this time from their positions in Libra and Aries. Unlike the comparatively gradual change represented by Virgo/Pisces, the Aries/Libra alignment will trigger sudden, unexpected events. But this is also likely to be the turning point from which a new paradigm will emerge.

This pairing will continue to affect business, politics, economics, individual freedoms, real estate, and health care. On a personal level it will manifest as profound change or endings. For some people the focus will be money matters, a relationship, or a job, for example. Others will experience Saturn/Uranus on an internal level that results in new directions. Be cautious. Changes made during a Uranus transit are almost always irreversible.

Saturn in Libra will also clash with Pluto in Capricorn in January and August, repeating the alignment that first occurred in November 2009. On a global scale, this represents a shrinking economy, workplace downsizing, and difficulties in the real-estate market. Frustration and an inability to control major forces of change are characteristic of this transit. On a personal level, this lineup can impact people's lives in any of these areas, while also encouraging a general desire and need to revamp and restructure life.

Alone, these planetary lineups are challenging. In combination, however, they represent a major structural shift that will occur this summer when Jupiter, Saturn, Uranus, and Pluto align in Aries, Libra, and Capricorn, all of which are cardinal, or action signs. This signals a swift and unexpected series of events in all of the above-mentioned areas. But with it will come the first hint of a fresh start, a glimpse into the future and unlimited opportunities to re-create life as we know it, both individually and collectively.

Ascendant Table

Your Time of Birth

Your Sun Sign	6–8 am	8–10 am	10 am–Noon	Noon–2 pm	2–4 pm	4–6 pm
Aries	Taurus	Gemini	Cancer	Leo	Virgo	Libra
Taurus	Gemini	Cancer	Leo	Virgo	Libra	Scorpio
Gemini	Cancer	Leo	Virgo	Libra	Scorpio	Sagittarius
Cancer	Leo	Virgo	Libra	Scorpio	Sagittarius	Capricorn
Leo	Virgo	Libra	Scorpio	Sagittarius	Capricorn	Aquarius
Virgo	Libra	Scorpio	Sagittarius	Capricorn	Aquarius	Pisces
Libra	Scorpio	Sagittarius	Capricorn	Aquarius	Pisces	Aries
Sagittarius	Sagittarius	Capricorn	Aquarius	Pisces	Aries	Taurus
Capricorn	Capricorn	Aquarius	Pisces	Aries	Taurus	Gemini
Aquarius	Aquarius	Pisces	Aries	Taurus	Gemini	Cancer
Pisces	Pisces	Aries	Taurus	Gemini	Cancer	Leo
	Aries	Taurus	Gemini	Cancer	Leo	Virgo

Ascendant Table

Your Time of Birth

Your Sun Sign	6–8 pm	8–10 pm	10 pm–Midnight	Midnight–2 am	2–4 am	4–6 am
Aries	Scorpio	Sagittarius	Capricorn	Aquarius	Pisces	Aries
Taurus	Sagittarius	Capricorn	Aquarius	Pisces	Aries	Taurus
Gemini	Capricorn	Aquarius	Pisces	Aries	Taurus	Gemini
Cancer	Aquarius	Pisces	Aries	Taurus	Gemini	Cancer
Leo	Pisces	Aries	Taurus	Gemini	Cancer	Leo
Virgo	Aries	Taurus	Gemini	Cancer	Leo	Virgo
Libra	Taurus	Gemini	Cancer	Leo	Virgo	Libra
Scorpio	Gemini	Cancer	Leo	Virgo	Libra	Scorpio
Sagittarius	Cancer	Leo	Virgo	Libra	Scorpio	Sagittarius
Capricorn	Leo	Virgo	Libra	Scorpio	Sagittarius	Capricorn
Aquarius	Virgo	Libra	Scorpio	Sagittarius	Capricorn	Aquarius
Pisces	Libra	Scorpio	Sagittarius	Capricorn	Aquarius	Pisces

How to use this table:
1. Find your Sun sign in the left column.
2. Find your approximate birth time in a vertical column.
3. Line up your Sun sign and birth time to find your Ascendant.

This table will give you an approximation of your Ascendant. If you feel that the sign listed as your Ascendant is incorrect, try the one either before or after the listed sign. It is difficult to determine your exact Ascendant without a complete natal chart.

Astrological Glossary

Air: One of the four basic elements. The air signs are Gemini, Libra, and Aquarius.

Angles: The four points of the chart that divide it into quadrants. The angles are sensitive areas that lend emphasis to planets located near them. These points are located on the cusps of the First, Fourth, Seventh, and Tenth Houses in a chart.

Ascendant: Rising sign. The degree of the zodiac on the eastern horizon at the time and place for which the horoscope is calculated. It can indicate the image or physical appearance you project to the world. The cusp of the First House.

Aspect: The angular relationship between planets, sensitive points, or house cusps in a horoscope. Lines drawn between the two points and the center of the chart, representing the Earth, form the angle of the aspect. Astrological aspects include conjunction (two points that are 0 degrees apart), opposition (two points, 180 degrees apart), square (two points, 90 degrees apart), sextile (two points, 60 degrees apart), and trine (two points, 120 degrees apart). Aspects can indicate harmony or challenge.

Cardinal Sign: One of the three qualities, or categories, that describe how a sign expresses itself. Aries, Cancer, Libra, and Capricorn are the cardinal signs, believed to initiate activity.

Chiron: Chiron is a comet traveling in orbit between Saturn and Uranus. Although research on its effect on natal charts is not yet complete, it is believed to represent a key or doorway, healing, ecology, and a bridge between traditional and modern methods.

Conjunction: An aspect or angle between two points in a chart where the two points are close enough so that the energies join. Can be considered either harmonious or challenging, depending on the planets involved and their placement.

Cusp: A dividing line between signs or houses in a chart.

Degree: Degree of arc. One of 360 divisions of a circle. The circle of the zodiac is divided into twelve astrological signs of 30 degrees each. Each degree is made up of 60 minutes, and each minute is made up of 60 seconds of zodiacal longitude.

Earth: One of the four basic elements. The earth signs are Taurus, Virgo, and Capricorn.

Eclipse: A solar eclipse is the full or partial covering of the Sun by the Moon (as viewed from Earth), and a lunar eclipse is the full or partial covering of the Moon by the Earth's own shadow.

Ecliptic: The Sun's apparent path around the Earth, which is actually the plane of the Earth's orbit extended out into space. The ecliptic forms the center of the zodiac.

Electional Astrology: A branch of astrology concerned with choosing the best time to initiate an activity.

Elements: The signs of the zodiac are divided into four groups of three zodiacal signs, each symbolized by one of the four elements of the ancients: fire, earth, air, and water. The element of a sign is said to express its essential nature.

Ephemeris: A listing of the Sun, Moon, and planets' positions and related information for astrological purposes.

Equinox: Equal night. The point in the Earth's orbit around the Sun at which the day and night are equal in length.

Feminine Signs: Each zodiac sign is either masculine or feminine. Earth signs (Taurus, Virgo, and Capricorn) and water signs (Cancer, Scorpio, and Pisces) are feminine.

Fire: One of the four basic elements. The fire signs are Aries, Leo, and Sagittarius.

Fixed Signs: Fixed is one of the three qualities, or categories, that describe how a sign expresses itself. The fixed signs are Taurus, Leo, Scorpio, and Aquarius. Fixed signs are said to be predisposed to existing patterns and somewhat resistant to change.

Hard Aspects: Hard aspects are those aspects in a chart that astrologers believe to represent difficulty or challenges. Among the hard aspects are the square, the opposition, and the conjunction (depending on which planets are conjunct).

Horizon: The word "horizon" is used in astrology in a manner similar to its common usage, except that only the eastern and western horizons are considered useful. The eastern horizon at the point of birth is the Ascendant, or First House cusp, of a natal chart, and the western horizon at the point of birth is the Descendant, or Seventh House cusp.

Houses: Division of the horoscope into twelve segments, beginning with the Ascendant. The dividing line between the houses are called house cusps. Each house corresponds to certain aspects of daily living, and is ruled by the astrological sign that governs the cusp, or dividing line between the house and the one previous.

Ingress: The point of entry of a planet into a sign.

Lagna: A term used in Hindu or Vedic astrology for Ascendant, the degree of the zodiac on the eastern horizon at the time of birth.

Masculine Signs: Each of the twelve signs of the zodiac is either "masculine" or "feminine." The fire signs (Aries, Leo, and Sagittarius) and the air signs (Gemini, Libra, and Aquarius) are masculine.

Midheaven: The highest point on the ecliptic, where it intersects the meridian that passes directly above the place for which the horoscope is cast; the southern point of the horoscope.

Midpoint: A point equally distant to two planets or house cusps. Midpoints are considered by some astrologers to be sensitive points in a person's chart.

Mundane Astrology: Mundane astrology is the branch of astrology generally concerned with political and economic events, and the nations involved in these events.

Mutable Signs: Mutable is one of the three qualities, or categories, that describe how a sign expresses itself. Mutable signs are Gemini, Virgo, Sagittarius, and Pisces. Mutable signs are said to be very adaptable and sometimes changeable.

Natal Chart: A person's birth chart. A natal chart is essentially a "snapshot" showing the placement of each of the planets at the exact time of a person's birth.

Node: The point where the planets cross the ecliptic, or the Earth's apparent path around the Sun. The North Node is the point where a planet moves northward, from the Earth's perspective, as it crosses the ecliptic; the South Node is where it moves south.

Opposition: Two points in a chart that are 180 degrees apart.

Orb: A small degree of margin used when calculating aspects in a chart. For example, although 180 degrees form an exact opposition, an astrologer might consider an aspect within 3 or 4 degrees on either side of 180 degrees to be an opposition, as the impact of the aspect can still be felt within this range. The less orb on an aspect, the stronger the aspect. Astrologers' opinions vary on how many degrees of orb to allow for each aspect.

Outer Planet: Uranus, Neptune, and Pluto are known as the outer planets. Because of their distance from the Sun, they take a long time to complete a single rotation. Everyone born within a few years on either side of a given date will have similar placements of these planets.

Planet: The planets used in astrology are Mercury, Venus, Mars, Jupiter, Saturn, Uranus, Neptune, and Pluto. For astrological purposes, the Sun and Moon are also considered planets. A natal or birth chart lists planetary placement at the moment of birth.

Planetary Rulership: The sign in which a planet is most harmoniously placed. Examples are the Sun in Leo, Jupiter in Sagittarius, and the Moon in Cancer.

Precession of Equinoxes: The gradual movement of the point of the Spring Equinox, located at 0 degrees Aries. This point marks the beginning of the tropical zodiac. The point moves slowly backward through the constellations of the zodiac, so that about every 2,000 years the equinox begins in an earlier constellation.

Qualities: In addition to categorizing the signs by element, astrologers place the twelve signs of the zodiac into three additional categories, or qualities: cardinal, mutable, or fixed. Each sign is considered to be a combination of its element and quality. Where the element of a sign describes its basic nature, the quality describes its mode of expression.

Retrograde Motion: Apparent backward motion of a planet. This is an illusion caused by the relative motion of the Earth and other planets in their elliptical orbits.

Sextile: Two points in a chart that are 60 degrees apart.

Sidereal Zodiac: Generally used by Hindu or Vedic astrologers. The sidereal zodiac is located where the constellations are actually positioned in the sky.

Soft Aspects: Soft aspects indicate good fortune or an easy relationship in the chart. Among the soft aspects are the trine, the sextile, and the conjunction (depending on which planets are conjunct each other).

Square: Two points in a chart that are 90 degrees apart.

Sun Sign: The sign of the zodiac in which the Sun is located at any given time.

Synodic Cycle: The time between conjunctions of two planets.

Trine: Two points in a chart that are 120 degrees apart.

Tropical Zodiac: The tropical zodiac begins at 0 degrees Aries, where the Sun is located during the Spring Equinox. This system is used by most Western astrologers and throughout this book.

Void-of-Course: A planet is void-of-course after it has made its last aspect within a sign, but before it has entered a new sign.

Water: One of the four basic elements. Water signs are Cancer, Scorpio, and Pisces.

Meanings of the Planets

The Sun

The Sun indicates the psychological bias that will dominate your actions. What you see, and why, is told in the reading for your Sun. The Sun also shows the basic energy patterns of your body and psyche. In many ways, the Sun is the dominant force in your horoscope and your life. Other influences, especially that of the Moon, may modify the Sun's influence, but nothing will cause you to depart very far from the basic solar pattern. Always keep in mind the basic influence of the Sun and remember all other influences must be interpreted in terms of it, especially insofar as they play a visible role in your life. You may think, dream, imagine, and hope a thousand things, according to your Moon and your other planets, but the Sun is what you are. To be your best self in terms of your Sun is to cause your energies to work along the path in which they will have maximum help from planetary vibrations.

The Moon

The Moon tells the desire of your life. When you know what you mean but can't verbalize it, it is your Moon that knows it and your Sun that can't say it. The wordless ecstasy, the mute sorrow, the secret dream, the esoteric picture of yourself that you can't get across to the world, or that the world doesn't comprehend or value—these are the products of the Moon. When you are misunderstood, it is your Moon nature, expressed imperfectly through the Sun sign, that feels betrayed. Things you know without thought—intuitions, hunches, instincts—are the products of the Moon. Modes of expression that you feel truly reflect your deepest self belong to the Moon: art, letters, creative work of any kind; sometimes love; sometimes business. Whatever you feel to be most deeply yourself is the product of your Moon and of the sign your Moon occupies at birth.

Mercury

Mercury is the sensory antenna of your horoscope. Its position by sign indicates your reactions to sights, sounds, odors, tastes, and touch impressions, affording a key to the attitude you have toward the physical world around you. Mercury is the messenger through which your physical body and brain (ruled by the Sun) and your inner nature (ruled by the Moon) are kept in contact with the outer world, which will appear to you according to the index of Mercury's position by sign in the horoscope. Mercury rules your rational mind.

Venus

Venus is the emotional antenna of your horoscope. Through Venus, impressions come to you from the outer world, to which you react emotionally. The position of Venus by sign at the time of your birth determines your attitude toward these experiences. As Mercury is the messenger linking sense impressions (sight, smell, etc.) to the basic nature of your Sun and Moon, so Venus is the messenger linking emotional impressions. If Venus is found in the same sign as the Sun, emotions gain importance in your life, and have a direct bearing on your actions. If Venus is in the same sign as the Moon, emotions bear directly on your inner nature, add self-confidence, make you sensitive to emotional impressions, and frequently indicate that you have more love in your heart than you are able to express. If Venus is in the same sign as Mercury, emotional impressions and sense impressions work together; you tend to idealize the world of the senses and sensualize the world of the emotions to interpret emotionally what you see and hear.

Mars

Mars is the energy principle in the horoscope. Its position indicates the channels into which energy will most easily be directed. It is the planet through which the activities of the Sun and the desires of the Moon express themselves in action. In the same sign as the Sun, Mars gives abundant energy, sometimes misdirected in temper, temperament, and quarrels. In the same sign as the Moon, it gives a great capacity to make use of the innermost aims, and to make the inner desires articulate and practical. In the same sign as Venus, it quickens emotional reactions and causes you to act on them, makes for ardor and passion in love, and fosters an earthly awareness of emotional realities.

Jupiter

Jupiter is the feeler for opportunity that you have out in the world. It passes along chances of a lifetime for consideration according to the basic nature of your Sun and Moon. Jupiter's sign position indicates the places where you will look for opportunity, the uses to which you wish to put it, and the capacity you have to react and profit by it. Jupiter is ordinarily and erroneously called the planet of luck. It is "luck" insofar as it is the index of opportunity, but your luck depends less on what comes to you than on what you do with what comes to you. In the same sign as the Sun or Moon, Jupiter gives a direct and generally effective response to opportunity and is likely to show forth at its "luckiest." If Jupiter is in the same sign as Mercury, sense impressions are interpreted opportunistically. If Jupiter is in the same sign as Venus, you interpret emotions in such a way as to turn them to your advantage; your feelings work harmoniously with the chances for progress that the world has to offer. If Jupiter is in the same sign as Mars, you follow opportunity with energy, dash, enthusiasm, and courage; take long chances; and play your cards wide open.

Saturn

Saturn indicates the direction that will be taken in life by the self-preservative principle that, in its highest manifestation, ceases to be purely defensive and becomes ambitious and aspiring. Your defense or attack against the world is shown by the sign position of Saturn in the horoscope of birth. If Saturn is in the same sign as the Sun or Moon, defense predominates, and there is danger of introversion. The farther Saturn is from the Sun, Moon, and Ascendant, the better for objectivity and extroversion. If Saturn is in the same sign as Mercury, there is a profound and serious reaction to sense impressions; this position generally accompanies a deep and efficient mind. If Saturn is in the same sign as Venus, a defensive attitude toward emotional experience makes for apparent coolness in love and difficulty with the emotions and human relations. If Saturn is in the same sign as Mars, confusion between defensive and aggressive urges can make a person indecisive. On the other hand, if the Sun and Moon are strong and the total personality well developed, a balanced, peaceful, and calm individual of sober judgment and moderate actions may be indicated. If Saturn is in the same sign as Jupiter, the reaction to opportunity is sober and balanced.

Uranus

Uranus in a general way relates to creativity, originality, or individuality, and its position by sign in the horoscope tells the direction in which you will seek to express yourself. In the same sign as Mercury or the Moon, Uranus suggests acute awareness, a quick reaction to sense impressions and experiences, or a hair-trigger mind. In the same sign as the Sun, it points to great nervous activity, a high-strung nature, and an original, creative, or eccentric personality. In the same sign as Mars, Uranus indicates high-speed activity, love of swift motion, and perhaps love of danger. In the same sign as Venus, it suggests an unusual reaction to emotional experience, idealism, sensuality, and original ideas about love and human relations. In the same sign as Saturn, Uranus points to good sense; this can be a practical, creative position, but, more often than not, it sets up a destructive conflict between practicality and originality that can result in a stalemate. In the same sign as Jupiter, Uranus makes opportunity, creates wealth and the means of getting it, and is conducive to the inventive, executive, and daring.

Neptune

Neptune relates to the deepest wells of the subconscious, inherited mentality, and spirituality, indicating what you take for granted in life. Neptune in the same sign as the Sun or Moon indicates that intuitions and hunches—or delusions—dominate; there is a need for rigidly holding to reality. In the same sign as Mercury, Neptune indicates sharp sensory perceptions, a sensitive and perhaps creative mind, and a quivering intensity of reaction to sensory experience. In the same sign as Venus, it reveals idealistic and romantic (or sentimental) reactions to emotional experience, as well as the danger of sensationalism and a love of strange pleasures. In the same sign as Mars, Neptune indicates energy and intuition that work together to make mastery of life—one of the signs of having angels (or devils) on your side. When in the same sign as Jupiter, Neptune describes an intuitive response to opportunity along practical and money-making lines. In the same sign as Saturn, Neptune indicates intuitive defense and attack on the world, which is generally successful unless Saturn is polarized on the negative side; then there is danger of unhappiness.

Pluto

Pluto is a planet of extremes—from the lowest criminal and violent level of our society to the heights people can attain when they realize their significance in the collectivity of humanity. Pluto also rules three important mysteries of life—sex, death, and rebirth—and links them to each other. One level of death symbolized by Pluto is the physical death of an individual, which occurs so that a person can be reborn into another body to further his or her spiritual development. On another level, individuals can experience a "death" of their old self when they realize the deeper significance of life; thus they become one of the "second born." In a natal horoscope, Pluto signifies our perspective on the world, our conscious and subconscious. Since so many of Pluto's qualities are centered on the deeper mysteries of life, the house position of Pluto, and aspects to it, can show you how to attain a deeper understanding of the importance of the spiritual in your life.

Forecasts

By Kris Brandt Riske

Aries . . . . Page 28
Taurus . . . . Page 51
Gemini . . . . Page 73
Cancer . . . . Page 96
Leo . . . . Page 119
Virgo . . . . Page 142
Libra . . . . Page 166
Scorpio . . . . Page 190
Sagittarius . . . . . Page 214
Capricorn . . . . . . Page 238
Aquarius . . . . . . Page 262
Pisces . . . . Page 286

ARIES

The Ram
March 20 to April 19

♈

Element: Fire
Quality: Cardinal
Polarity: Yang/Masculine
Planetary Ruler: Mars
Meditation: I build upon my strengths
Gemstone: Diamond
Power Stones: Bloodstone, carnelian, ruby
Key Phrase: I am

Glyph: Ram's head
Anatomy: Head, face, throat
Color: Red, white
Animal: Ram
Myths/Legends: Artemis, Jason and the Golden Fleece
House: First
Opposite Sign: Libra
Flower: Geranium
Key Word: Initiative

Your Strengths and Challenges

Your sign, the first in the zodiac, is noted for action, energy, and initiative. You're often first in line, leading the way and forging ahead into new territory. Persistence is a strength, but impatience can get the best of you when a more thoughtful approach would produce better results. Bold and daring, your aura is fresh and new, with a charming innocence that makes you eminently approachable and endears you to many.

Fiery Mars, your ruling planet, inspires you to excel in every activity. The red planet also encourages you to take charge and get things moving. You're the first to accept a challenge, and that's a plus as long as you combine these traits with teamwork and cooperation.

You live in the here and now, which works well most of the time. But remember that you can learn from the past, and you'll get further in life if you plan for the future. The same is true in decision-making. Although quick action has its advantages, snap decisions can backfire. Consider the potential consequences, the upside and the downside, before you act, and learn the difference between impulsiveness and dynamic action.

People are amazed by your energy. Sports and exercise, which most Aries enjoy, are a great way to relax, and playtime is essential to your well-being. Physical activity also helps free your mind and your sixth sense. The more you listen to your inner voice, the greater the chance to develop your strong intuitive potential as well as deepen your self-knowledge.

Your Relationships

Friends are near and dear to you, although you have far more acquaintances than close pals. You have a wide circle of people who are in some way unusual—career, personality, interests, hobbies. You're also a natural networker, and enjoy bringing people together both for their benefit and your own. Many born under your sign are leaders in clubs and organizations.

Your love life is distinguished by energetic enthusiasm. Playful and romantic, you know how to keep love alive, and instinctively create the ideal atmosphere for romance. At times your love life is as dramatic as a soap opera, but pride can make it tough to let go even when you know that's best. Eventually, though, you do, and love blooms anew while playing the field. Although Aries is a self-reliant sign, you place a high value on partnership and are motivated to find a mate. You could make a sensational match with a Libra, your opposite sign, who can teach you a lot about people. Love could sizzle with one of the other fire signs, Leo and Sagittarius, but you might find it tough to coordinate schedules with another Aries; and Cancer and Capricorn may not be adventuresome enough to suit your taste.

Most people born under your sign have a strong need for children, although you prefer a small family to a large one. Overindulgence can stem from your desire to provide your children with the best of everything. But keep in mind that children need age-appropriate responsibilities in order to develop solid values and ethics. Be more generous with your time than your money, and encourage them to develop their interests and talents.

Your Career and Money

You're a dynamic go-getter with an eye on moving up the ladder. Here, too, impatience can stall progress until you fully realize that each step requires time and effort. Do that with a practical approach, and you can be among the most successful in your career field. You're the one others depend on to come through, time after time. But draw the line at taking on more tasks than you can reasonably handle, as well as picking up the slack for others. You're meticulous at work and also excel at analysis, which gives you the ability to see what others miss. You have excellent earning potential and can build sizeable assets with long-term investments. Most Aries are frugal, and you have an eye for value, spending only when you get your money's worth. But guard against a tendency to take big financial risks that could defeat your ultimate goal of financial security.

Your Lighter Side

Spring arrives every year when the Sun enters your sign, and you carry that season's energy with you all year long, wherever you go. Your presence sparks passion. People find your charming can-do attitude irresistible, and your motivational talent is unmatched. It's the perfect formula to throw people and projects into high gear.

Affirmation for the Year
Every challenge is ultimately an opportunity!

The Year Ahead for Aries

Jupiter wraps up its year-long influence on friendship, socializing, and group activities when it exits Aquarius, your solar Eleventh House, January 17. Although Jupiter's tour of Pisces won't have the dynamic outward energy associated with Aquarius, the lucky planet is even luckier as it transits your solar Twelfth House.

Jupiter in the Twelfth House functions much like a guardian angel, offering an extra level of protection and good fortune. It's the hidden luck factor that can come to your rescue at the eleventh hour. Just don't expect it to deliver 100 percent of the time because Jupiter is also known for "promises, promises!"

Any planet transiting the Twelfth House signals a period of self-renewal and preparation for new beginnings. So you should take time to look inward and assess your life. During the process a deeper level of spirituality may emerge or you might strive for a greater degree of self-understanding in a quest to learn what motivates you and what holds you back. Use the time wisely and you'll benefit from the knowledge in 2011. Even better, you'll get a preview when Jupiter temporarily visits your sign June 5 through September 7. Think of this as bonus—a time to catch a glimpse of the Jupiterian bounty that can come your way next year. Make your own luck; seize the opportunity to explore new personal territory that can come full circle in 2011. Early June is a particularly lucky period, especially if you were born within the first few days of your sign.

Change has been a constant the past few years with Saturn in Virgo, your solar Sixth House, at odds with Uranus in Pisces. You've undoubtedly experienced an upheaval in your work life, whether personally or in your work environment. Although this influence has mostly run its course, Saturn will complete its Virgo transit between April 7 and July 20, and make its last Virgo/Pisces contact with Uranus the end of April. Be aware and be proactive if you see further changes coming at work. If you're job-hunting (or ready for a more rewarding position) you could realize success during that time frame.

Saturn begins and ends the year in Libra, your solar Seventh House, with time out for its brief retreat into Virgo. Here, too, it will clash with Uranus as that planet advances into your sign May

27, before returning to Pisces August 13. So you can expect change in close relationships, not just this year but off and on during the next three. Don't be surprised if you purposely distance yourself from some people, whether friends, relatives, or coworkers. But other ties will strengthen and deepen, and you can learn much from people who come into your life during this time.

With Uranus's short visit in Aries comes your first clue as to how this planet of change and the unexpected might manifest in your life during the next seven years. This influence will bring a feeling of dissatisfaction with the status quo and the urge for change. That's great if your goal is self-improvement, but a major life change is probably premature at this point. Use this year instead to imagine and explore possibilities; expect them to evolve with time. Besides, you'll want to focus as much, or possibly more, of your attention on Uranus's Pisces transit, which runs through March 2011. Placed here, this planet is an asset if you could benefit from an overall healthier lifestyle, regular exercise, and a more nutritious diet. Uranus offers you the strength for change and its alignment with Saturn encourages determination. But do yourself a favor and first get a checkup. If stress is an issue, find a new leisure-time activity that yields creative yet tangible results, such as gardening, woodworking, or home improvements.

Neptune continues its multi-year transit of Aquarius, the sign it entered in 1998. Placed in your solar Eleventh House, this mystical planet encourages you to inspire others, whether one-on-one with friends or in groups and organizations. You can be the motivator, the one who helps bring out the best in people for all the right reasons. But Neptune is also the planet of disillusionment, so take care not to put too much faith in people; after all, they're only human, and unrealistic hopes can result in disappointment.

Pluto, now firmly established in Capricorn, spotlights your solar Tenth House of career. This powerful placement, which lasts until 2024, can help you achieve ambitious goals. Like all planets, Pluto also has a potential downside. Here, it can trigger work-related personality clashes, downsizing, and excessive rules and regulations that interfere with job activities and advancement. Because of Pluto's long transit, you might experience all this on a minor level most years and to a greater extent only when Pluto contacts your Sun.

Nevertheless, with Pluto active in your career sector, you should make it a point to be aware of what's happening in your workplace and your career field, particularly this summer, when Pluto will form a difficult alignment with Jupiter, Saturn, and Uranus, undoubtedly prompting workplace and/or relationship changes.

What This Year's Eclipses Mean for You

Eclipses typically have a six-month influence, when their energy is periodically activated. With two of this year's eclipses in Capricorn, your solar Tenth House, career matters will be a main focus throughout 2010. The other two are in Cancer, your solar Fourth House, and Gemini, your solar Third House.

The January 15 solar eclipse in Capricorn launches 2010 with an emphasis on your career, which is renewed at the June 26 lunar eclipse in the same sign. Although similar, each is different because each is aligned with different planets. This pair of eclipses could trigger a new job or promotion.

You and your career will benefit most from the January 15 solar eclipse in Capricorn because of its alignment with Venus, also in Capricorn. Chances are, you'll be among the favored few career-wise, and well placed for a step up in the world if that's your desire.

The June 26 lunar eclipse in Capricorn also highlights your career sector, but with an additional emphasis on home and family relationships. Your challenge with this eclipse will be to balance career and domestic life. Communication is key, and loved ones will be far more supportive during this potentially stressful career time if you share your thoughts and feelings and make family time a priority.

Domestic life will also be in the forefront as the July 11 solar eclipse in Cancer spotlights your solar Fourth House. Although relocation is a possibility with this influence, the odds favor exciting, upbeat family activities and do-it-yourself projects. Tap your creativity and make your place a showplace.

December 21 brings a lunar eclipse in Gemini, your solar Third House of communication and quick trips. Aligned as it is with Mercury in Sagittarius, your solar Ninth House, you can expect more frequent travel in the first six months of 2011, some of it unexpected. This eclipse also emphasizes education. Consider taking one class or several to benefit your career.

Saturn

If you were born between April 17 and 19, Saturn will contact your Sun from Virgo, your solar Sixth House of daily work, between April 7 and July 20. This coincides with the April 26 Saturn/Uranus alignment, so you'll want to be especially alert to subtle workplace signs and signals—casual comments, business activity, and any developing trends in your career field. Downsizing is possible with this planetary influence, which could affect your position either directly or indirectly. If the latter, it might be necessary to take on additional responsibilities. Think back to the fall of 2009, when this planetary alignment last occurred; that may give you a clue about what might occur this spring or summer.

Virgo is also your health and wellness sign, so you should schedule routine medical, dental, and eye checkups, and learn stress-management techniques. Because Saturn represents permanence and Uranus represents change, you can use this dual energy to replace long-standing habits with positive ones and to revamp your diet and exercise program (or begin one!) to improve your overall health. Another excellent outlet for this Saturn/Uranus lineup is volunteer activities, whether through an ongoing commitment to a local organization or participation in a one-time event such as a charity walkathon.

If you were born between March 20 and April 16, Saturn in Libra, your solar Seventh House of relationships, will contact your Sun at some point during 2010. Saturn's influence will be minor for most, no more than a week or two during the final five months of the year. Some relationships will be strained and you may feel as though you've been singled out to carry the full load while others do little or nothing. You may also tire easily and choose to spend more time alone. All this is typical of Saturn, which encourages you to reflect upon the people in your life and what you can learn from them.

If your birthday is between March 20 and 25, however, Saturn will make two contacts, first between January 1 and April 6, and again from late July to early August. You'll thus experience ups and downs in at least some relationships (family, partner, coworkers, close friends), and any related events that occur during the first three months of the year will be resolved, one way or another, this summer. Some people will enter your life, some will depart, and some ties will be strengthened. Someone from the past could reenter your life during either of these time frames, and if you're debating the merits of a commitment, your thoughts and feelings will crystallize in July.

Treat yourself especially well during the first part of the year. Saturn's contact with your Sun can result in reduced vitality, so make sleep, rest, and leisure time a priority. This, combined with healthy eating, will help promote tolerance, a quality you'll definitely need when others frustrate you. Try to remember, though, that it's not so much them as it is you and your perception. Listen, emphasize patience, and turn even the most trying exchanges into valuable learning experiences about yourself, relationships, and human nature.

Uranus

Uranus in Aries briefly joins forces with your Sun **if you were born between March 20 and 22**. During its time in Aries (May 27 to August 12) you're likely to feel a higher level of restlessness and an increasing urge for change. Both effects will wane somewhat as Uranus slips back into Pisces for the rest of the year, giving you time to think. So now is not the time to make snap decisions or to initiate sudden moves; they're likely to be undone next year when Uranus returns to your sign.

Now is the time, however, for an honest self-assessment. Are you happy? Is your career fulfilling? What can you do to improve a personal or professional partnership? Do you feel iffy about a potential partnership, and if so, why? Thoughtful meditation could yield surprising insights and answers to these questions and more, so take an honest look at yourself and your life as a first step toward becoming the person you want to be.

Neptune

If you were born between April 13 and 19, Neptune in Aquarius forms a favorable contact with your Sun. Placed in your solar Eleventh House, Neptune offers you the opportunity to inspire others, especially friends, and to share your skills and talents in group endeavors, including good deeds that will have you feeling warm and fuzzy. You can be a marvelous team player this year, the motivational sparkplug that brings people together to achieve a common goal.

But you'll also want to be a bit skeptical of anyone new who comes into your life. Despite being an overall positive influence, Neptune can color thinking and cloud your vision, encouraging you to see only the best in others. So be cautious and don't put your faith in just anyone or just any cause, no matter how much the message touches your mind and heart.

Neptune also encourages creativity, and this is a great year to develop a talent even if you've never ventured into this realm. Creativity, however, isn't limited to artistic pursuits; hobbies and other leisure-time activities can awaken this energy, which will also be an asset in problem-solving and brainstorming.

Listen to your inner voice this year. Neptune will open your intuitive pathway, and this can give you an edge in personal and workplace relationships. On another level, Neptune urges you to explore lifetime goals and ambitions in order to better define your place in the world, what you value, and how best to express your individuality. Dream big!

Pluto

Pluto will contact your Sun from Capricorn, your solar Tenth house, **if you were born between March 23 and 26.** Keep your wits about you and stay in tune with what's happening in your job and career field. You could be involved in significant company downsizing this year, if not in January, then this summer. In any case, you should prepare yourself for major career changes from this transit. Although difficult at the time, you will appreciate the value of this Pluto-Sun contact in a year or so. Your life and your career will ultimately be the better for the transformation that occurs.

Jupiter-Saturn-Uranus-Pluto

If you were born between March 20 and 24, this year's Jupiter-Saturn-Uranus-Pluto alignment will contact your Sun this summer. Unfortunately, there is nothing easy about this unusual four-planet lineup that forecasts large-scale change on both a personal and global scale. For you, it involves your Sun and solar First House

(Jupiter, Uranus), Seventh House (Saturn), and Tenth House (Pluto)—self, relationships, and career.

There are a number of ways in which this alignment can impact your life, including difficulties in personal and workplace relationships, with one affecting the other. You could find yourself caught in a power play at work, or become jobless through no fault of your own. Although this is likely to be stressful, the planetary contact with your Sun presents an opportunity to reinvent yourself and your place in the world. This, however, can strain a close personal relationship so it's important to keep your partner in the loop. But be cautious and patient if a committed relationship approaches the breaking point; premature action can lead to major regrets. The same applies if you want to launch a business or personal partnership. This is not the year to become an entrepreneur, even though expansive, upbeat Jupiter encourages new ventures.

Your best path this year is to keep your options open and to steer clear of workplace conflict while being alert to any career undercurrents. Most of all, think before you act and be true to your values and ethics rather than succumb to peer pressure. Ultimately, you may decide to pursue a new career direction, if not this year then next.

This planetary alignment also has a potential upside. With Jupiter meeting Uranus in your sign, you'll be in a lucky phase this summer, one that can trigger a sudden opportunity (or turn a minor spill into a major mess). Even so, it won't be an effortless process, and to some extent you'll need to make your own luck; initiative and follow-through are all-important. Much depends upon the actions you take and the people you side with. Wrong choices here can backfire—big time—which makes careful thought and observation a must before you seek anyone's support or snap up an opportunity. Even then, the first chapter is not the last because the events of 2010 will continue to evolve in 2011.

Aries/January

Planetary Hotspots
Career relationships are this month's hotspot as the planetary energy builds toward a Saturn/Pluto clash in your solar Seventh and Tenth Houses. Be alert in the days surrounding the 15th, when Mercury turns direct in Capricorn. You may get hint of what's to come, which could be a workplace shake-up. And, be cautious what you say to whom; a friendly coworker may be a hidden foe.

Planetary Lightspots
You're in high focus, popular and among the favored few in early January when the Sun, Mercury, and Venus merge in Capricorn. Snap up any chance to talk with people, impress decision-makers, make a presentation, and network. What you do then could bring a pleasant surprise midmonth.

Relationships
Friendship is in the spotlight after Venus enters Aquarius, your solar Eleventh House, on the 18th, followed by the Sun on the 19th. Fill your calendar with events, but skip the last few days of January, when Venus and the Sun will clash with Mars. That lineup and the Leo Full Moon on the 30th could signal the end of a dating relationship—or the beginning of sizzling romance.

Money and Success
Handle controversy, conflict, and controlling people with finesse and you could be in line for rewards. At the least you'll hear praise the third week of January. Risky investments are not the way to go with Mars retrograde in Leo, your solar Fifth House, and you may be reminded of just how expensive children can be. Be patient and spoil them with time and attention instead of money.

Rewarding Days
1, 2, 4, 8, 11, 14, 17, 22, 25

Challenging Days
6, 13, 16, 20, 21, 23, 27, 30

Aries/February

Planetary Hotspots
Friendship is a potential February hotspot. Not that you won't enjoy your pals and socializing with them. Far from it. But be a little wary, especially of anyone new. The Aquarius New Moon on the 15th in your solar Eleventh House aligns with Neptune in the same sign, and Mercury does the same near month's end. Someone may be less than truthful with you and make lofty promises with no intention of keeping them. Be skeptical.

Planetary Lightspots
Jupiter in Pisces, your solar Twelfth House guardian angel, will be active at month's end when it meets the Sun in the same sign. Your career is likely to be the main beneficiary, when someone behind the scenes promotes your talents. Don't hesitate to ask for a favor; chances are, it will be granted.

Relationships
Misunderstandings are possible midmonth, when Mars, still retrograde in Leo, aligns with Saturn in Libra, your solar Seventh House. At work, the issue could be a stalled project; at home, you and your mate may disagree over how best to handle one of your children. Keep cool and listen before forming an opinion.

Money and Success
Although retrograde Mars continues to advise against investments and related decisions, the timing is great for organizing financial records and gathering tax information. While you're at it, ask for updated quotes on insurance, and read the fine print regarding employee benefits. You could net a cost savings as a result. Just postpone final decisions until at least mid-March, when Mars will turn direct.

Rewarding Days
3, 6, 7, 8, 13, 16, 18, 26, 27

Challenging Days
2, 5, 9, 12, 17, 21, 23

Aries/March

Planetary Hotspots
January's career-related stresses and strains return to life this month as several planets activate Saturn in Libra, your solar Seventh House, and Pluto in Capricorn, your solar Tenth House. Keeping a low profile will be tough this time, so mentally prepare yourself to go with the flow. And be sure you deliver every project on time and in top shape. Exceed expectations.

Planetary Lightspots
You have two planetary events to celebrate this month. The first is Venus, which enters your sign March 7. You can charm almost anyone you choose, and your powers of attraction will be at their best. Share your dreams with the universe and expect them to come true. The second is Mars, your ruler, which resumes direct motion in Leo on the 10th. Although it may take a few weeks to regain full momentum, you'll feel like your life is again moving forward.

Relationships
Someone from the past could reappear at the least expected moment. As tempted as you might be to rekindle the tie, you may soon regret it. Remember why you parted ways. A friend gives you wise advice at month's end. Listen closely. A shaky relationship could end near the Full Moon in Libra on the 29th when ideas and opinions clash, but some singles launch a whirlwind romance in early March.

Money and Success
Found money could be in your near future, thanks to a Venus-Uranus merger in Pisces on the 3rd. Even better, Venus enters Taurus, your solar Second House, on the 31st, to begin a positive three-week financial trend.

Rewarding Days
4, 7, 9, 10, 12, 13, 17, 19, 22, 26, 30

Challenging Days
2, 8, 11, 16, 18, 20, 23, 24, 25, 29, 31

Aries/April

Planetary Hotspots
Saturn returns to Virgo, your solar Sixth House, on the 7th, and forms an exact alignment with Uranus in Pisces on the 26th. This makes your work life the April hotspot. A coworker or supervisor could push you to the limit, and it's also possible that downsizing could affect your job, either directly or indirectly. If your job is secure, there may be a change in benefits or expectations, or a new boss could prompt you to look elsewhere. Hang in there! Find a stress reliever.

Planetary Lightspots
You're in the limelight, thanks to the April 14 New Moon in your sign. Take this fresh energy with you every day and step out with ultimate confidence. In quieter moments give some thought to what you want to accomplish in the next twelve months. Dream big!

Relationships
You'll experience just about every facet of relationships this month, from aggravating to upbeat. Your patience will be tested at times, particularly early and late in the month, and a friend could provide the inspiration you need to launch your new solar year with the highest of hopes. Listen carefully when someone offers you the benefit of knowledge and experience. This wise person can see what is invisible to you.

Money and Success
Mercury turns retrograde on the 17th in Taurus, your solar Second House of personal resources. Safeguard valuables when you're out in public, and pay bills early just in case a mix-up should occur. If possible, put major financial decisions and purchases on hold until after Mercury resumes direct motion May 11. Mercury's retrograde period is excellent, however, for finding bargains on clothing.

Rewarding Days
2, 3, 8, 9, 13, 14, 17, 21, 26, 27, 30

Challenging Days
4, 7, 12, 19, 23, 25, 28

Aries/May

Planetary Hotspots

A big event occurs on the 27th: Uranus enters your sign. Although it probably won't be prominent in your life until June or July, you will begin to feel the restless urge that accompanies this transit. Pay attention, and even consider jotting down your thoughts when the desire for change surfaces. This initial, albeit subtle information can help you zero in on the new directions that are in your future.

Planetary Lightspots

Need a break? Plan a weekend getaway or a vacation trip around the time of the May 27 Full Moon in Sagittarius, your travel sign. If that's impossible, take a day trip, or stock up on library books to satisfy your desire for knowledge and adventure.

Relationships

You'll be in sync with almost everyone, with Venus in Gemini, your solar Third House of communication, through the 18th. From there, the Sun takes over, entering the same sign on the 20th. Use this month to touch base with relatives, meet neighbors, and mend fences. If you're a parent, plan a special outing with your children the weekend of the 7th.

Money and Success

The May 13 New Moon in Taurus energizes your solar Second House. A raise or bonus, an extra perk, or a promotion is possible. The outcome is linked to an exact alignment of Saturn and Jupiter across your solar Sixth/Twelfth House axis. If you're satisfied with your job, you'll welcome the boost provided by this planetary lineup. But the duo can also trigger inner conflict if the urge for freedom clashes with your sense of responsibility. Think first about the long-term effects; then make a decision.

Rewarding Days

1, 5, 6, 10, 11, 14, 15, 19, 24, 27, 28

Challenging Days

2, 9, 16, 18, 22, 23, 25, 29, 30

Aries/June

Planetary Hotspots
Jupiter enters your sign on the 6th, and joins forces with Uranus on the 8th. Translation: pure luck! So why is this a hotspot? It may or may not be, depending upon how things play out when both planets clash with Pluto in Capricorn, your solar Tenth House, around the time of the June 26 Full Moon eclipse in the same sign. That could bring difficulties with a supervisor, or promotion to a top spot. If the latter, get all the facts because the opportunity may look far rosier than it really is. Think before you act. Matters will continue to develop into July and August.

Planetary Lightspots
Socializing and romance are in the forecast after Venus arrives in Leo, your solar Fifth House, on the 14th. Venus aligns beautifully with several planets, increasing the odds for an active love life or a new love interest. You could network your way into career success at a social event.

Relationships
Family communication will require the ultimate in patience and adaptability as several planets in Cancer, your solar Fourth House, clash with others. At issue may be family time versus career time, and it will be a balancing act to keep up with both and keep everyone happy, including yourself. Get family input if you're contemplating a big-ticket purchase or home improvement.

Money and Success
Mars dashes into Virgo, your solar Sixth House, on the 7th. Expect the pace to pick up at work, and keep in mind that what you do this month will directly impact events later this summer. Keep controversial thoughts to yourself the week of the 12th.

Rewarding Days
1, 11, 15, 16, 17, 20, 23, 24, 28, 29

Challenging Days
2, 5, 6, 12, 13, 19, 25, 26

Aries/July

Planetary Hotspots
Saturn returns to Libra, your solar Seventh House, on the 21st, and forms an exact alignment with Uranus in Aries on the 26th. Change will surround a long-term relationship (personal or professional) that could reach the breaking point. It will be tough to agree to disagree now, so time apart might be a good option. Only you can know what's right for you, but keep in mind that you could regret a snap decision.

Planetary Lightspots
The July 11 New Moon eclipse spotlights your domestic sector, promising happy family times and relationships. You're drawn to home life and motivated to get your place in shape. Get moving while you're in the mood and dive into do-it-yourself projects. Then show off the results by hosting a get-together for friends and family.

Relationships
Plan ahead. Your social life will keep you on the go this month, thanks to the Sun, Mercury, and Venus in Leo, your solar Fifth House; and the July 25 Full Moon in Aquarius, your solar Eleventh House of friendship. You'll also cherish special moments with your children, and romantic ones with your partner or a date. Again this month, you could make a fabulous career connection while socializing with friends.

Money and Success
You're nicely placed to reap career rewards. The level of success depends of course upon performance and also the people you know. Lucky connections can help you rise to the top, and you'll want to showcase your talents at every opportunity. Lead by example, share the credit and your knowledge, and be a motivational force.

Rewarding Days
2, 4, 8, 12, 13, 17, 21, 25, 27, 31

Challenging Days
3, 9, 10, 16, 19, 20, 22, 23, 30

Aries/August

Planetary Hotspots
Events that began in June or July near a conclusion as Jupiter in Aries aligns with Saturn in Libra and Pluto in Capricorn. Career developments are on tap, and relationships will have an integral role in the outcome. Whatever occurs, events signal a new opportunity on the horizon. A professional (or personal) relationship may end, but this is not the time to launch a new business endeavor. Take note of events early in the month; they may give you insight into what will occur the third or fourth week of August.

Planetary Lightspots
The August 24 Full Moon in Pisces, your solar Twelfth House, encourages you to take some much-needed time out for yourself. Catch up on sleep, and fill your non-working hours with relaxing activities. Dreams could be particularly insightful, and meditation can help your subconscious thoughts and feelings emerge.

Relationships
Your social life continues to be active the first three weeks of August as the Sun advances in Leo, your solar Fifth House. The energy peaks the week of the New Moon in the same sign on the 9th. Plan a social event with friends that weekend; if you're single, someone intriguing could catch your eye. Time with your children is good for your soul and theirs. A friend could disappoint you around the 20th.

Money and Success
Mercury turns retrograde in Virgo, your solar Sixth House of daily work, on the 20th. Take time to double-check details even if you're rushed, and confirm all instructions before beginning a new task. This is also your wellness sector, so take precautions to avoid a cold or flu.

Rewarding Days
1, 4, 9, 11, 13, 14, 17, 22, 23

Challenging Days
2, 3, 5, 6, 10, 12, 18, 19, 27

Aries/September

Planetary Hotspots
The September 23 Full Moon in your sign triggers what may be the final chapter in this summer's events involving you, those close to you, and your career. Job and relationship changes are again possible, and the lunar energy provides the incentive and initiative to set and pursue new personal and professional goals. Some long-term relationships may end, by choice or circumstance, mostly because these people have fulfilled their role in your life.

Planetary Lightspots
Jupiter returns to Pisces, your solar Twelfth House, on the 8th, and joins forces with Uranus on the 18th. This lineup puts luck on your side and also encourages you to have faith in the universe and yourself. Networking can be very successful now.

Relationships
The communication difficulties of the past few weeks begin to clear up as Mercury resumes direct motion on the 12th, and people will welcome your desire to clear up misunderstandings. Listen closely to the wise words of a friend even if it isn't what you want to hear. Open your mind, accept the reality, and then let the information inspire you to embrace the future.

Money and Success
Venus enters Scorpio, your solar Third House, on the 8th, followed by Mars on the 14th. This lineup is promising for your bank balance, but save rather than spend your gains. Around the 21st, an unexpected job development could ultimately be in your favor, so be sure to consider any opportunity that pops up that week.

Rewarding Days
2, 4, 5, 6, 10, 14, 17, 19, 24, 28

Challenging Days
1, 3, 7, 9, 15, 23, 25, 30

Aries/October

Planetary Hotspots
Venus turns retrograde in Scorpio, your solar Eighth House, on the 8th, which can slow family income during the next six weeks. Besides paying bills early, you can take advantage of this time to review budgets and investments, and to analyze both short- and long-term financial goals. But hold off on major money decisions until the middle or end of November. Month's end is favorable for family financial discussions, and with a little effort you can reach a workable compromise that satisfies everyone's needs.

Planetary Lightspots
Look forward to the 29th, the date of the Full Moon in your sign. It has the potential to energize you and your life with enthusiasm. This lunar energy aligns with Neptune in Aquarius, encouraging you to dream about and then initiate the future. It also boosts your sixth sense, so listen when your subconscious wishes emerge.

Relationships
You'll be in tune with almost everyone under the October 7 New Moon in Libra, which accents your solar Seventh House, and prompts some to take a relationship to the next level. But postpone commitment until Venus turns direct. You also can gain valuable and practical advice from professional consultations, and be successful in gaining support for your endeavors. This can be a magical time for couples and their mutual goals.

Money and Success
Although Venus is retrograde in Scorpio, favorable planetary alignments in this sector signal financial luck and progress toward increasing your income. Just be aware that it may be a couple of months before what you do now yields results.

Rewarding Days
7, 8, 9, 11, 15, 17, 25, 26, 29

Challenging Days
2, 3, 6, 13, 14, 20, 21, 23, 27

Aries/November

Planetary Hotspots
Holiday travel isn't the best idea this year. Challenging planetary alignments that week could trigger a delay, cancellation, lost luggage, or all three. Even if they don't, time with friends or relatives will be less than enjoyable despite your enthusiasm. Consider, instead, inviting a few people to your place for a late afternoon potluck Thanksgiving dinner, followed by time for your interests.

Planetary Lightspots
Learning is a strong theme this month with Mercury in Sagittarius, your solar Ninth House, from the 18th to the 29th. Take a quickie class for the fun of it, read for relaxation, or use the Internet as a source for how-to skills. You might also consider volunteering for a good cause because this transit touches your spiritual side.

Relationships
Venus resumes direct motion in Libra, your solar Seventh House, on the 18th. Although it won't happen instantly, you can look forward to enhanced communication with loved ones during the next few weeks. Once Venus turns direct you'll also find people more receptive to your requests. If you're contemplating a romantic commitment, month's end brings clarity.

Money and Success
Both the New Moon (November 6) in Scorpio and the Full Moon (November 21) in Taurus have the potential to boost your income through a raise, bonus, or minor windfall. At the least you can expect money matters to be fairly stable. You might also make a lucky find or cash in on a group lottery win midmonth. If you're in the market for a loan or mortgage, the same timing could bring success; but borrow less rather than more.

Rewarding Days
4, 7, 8, 12, 13, 16, 18, 22, 26

Challenging Days
3, 9, 17, 23, 24, 25, 27, 30

Aries/December

Planetary Hotspots

Mercury turns retrograde in Capricorn, your solar Tenth House, on the 10th, which can trigger career challenges, mix-ups, and misunderstandings. Ask for clarification rather than to assume instructions are clear, and also take time to review all work. The last thing you want now is to give anyone the opportunity to criticize your output. Also expect decisions to be delayed. Be patient.

Planetary Lightspots

The December 5 New Moon in Sagittarius, your solar Ninth House, emphasizes travel. If fresh scenery is your goal, go the second week of December rather than over the holidays. Then you'll get the best of the planetary alignments and increase the odds for a carefree trip. You'll also avoid retrograde Mercury, which retreats in Sagittarius on the 18th, increasing the potential for travel disruptions.

Relationships

Relationships benefit from open communication, and the December 21 Full Moon (lunar eclipse) in Gemini brings that opportunity. You'll also have increased contact with neighbors, and involvement in a community activity could benefit from your leadership skills. Although contact with relatives is generally positive now, rising tension could prompt you to finally resolve long-standing issues.

Money and Success

Your career gets the best—and unfortunately also the worst—of several planets in Capricorn, your solar Tenth House. Success is indicated in early December, but you'll have to contend with difficult, controlling people as the year comes to a close. Frustration will rise as someone blocks progress. Consider the source, accept it for what it is, and find another outlet to vent your frustrations.

Rewarding Days
4, 5, 10, 16, 19, 23, 25, 29, 31

Challenging Days
7, 8, 11, 14, 15, 20, 21, 27, 28

Aries Action Table

These dates reflect the best—but not the only—times for success and ease in these activities, according to your Sun sign.

	JAN	FEB	MAR	APR	MAY	JUN	JUL	AUG	SEP	OCT	NOV	DEC
Move					21, 22	4, 5, 9, 10, 25–30	1–9					
Start a class	25, 26	22, 23				10–20						
Join a club	20–31	1–17										
Ask for a raise		19		15, 16	12, 31							
Look for work	4, 5				21, 22		14, 15, 28–31	1–6	7, 8, 16–22			
Get pro advice	6, 7	3, 4										
Get a loan			4, 5	1, 2					11			2, 3
See a doctor		24, 25	4, 5					11	20			
Start a diet		28	1							5, 6	1, 2	
End relationship	6, 7	3, 4	2, 3, 29, 30								3, 4, 30	1
Buy clothes			25, 26			15, 16		9, 10			26, 27	
Get a makeover	21, 22		11, 12, 21–30	12–14				10				
New romance		26, 27	25, 26			15, 16	12					
Vacation											22–30	

TAURUS

The Bull
April 19 to May 20

☒

Element: Earth
Quality: Fixed
Polarity: Yin/Feminine
Planetary Ruler: Venus
Meditation: I trust myself and others
Gemstone: Emerald
Power Stones: Diamond, blue lace agate, rose quartz
Key Phrase: I have

Glyph: Bull's head
Anatomy: Throat, neck
Color: Green
Animal: Cattle
Myths/Legends: Isis and Osiris, Cerridwen, Bull of Minos
House: Second
Opposite Sign: Scorpio
Flower: Violet
Key Word: Conservation

Your Strengths and Challenges

Determination, willpower, and follow-through are traits that set you apart from the crowd. Taken too far, however, these positives can manifest as stubbornness. Stand your ground on the important issues and learn not to resist the rest. Your steady, practical nature is also a strength, and although you do everything at your own pace, you almost always come through, sometimes ahead of others. This stabilizing energy also makes people feel secure in your presence, and those closest to you value the "safe haven" aura you project.

Taurus finds security in routine, which makes change difficult. But sometimes change can be good; otherwise, it's easy to slip into a rut, which can limit your opportunities. So take the initiative to stir things up once in a while! Start small to raise your comfort level.

With Venus as your ruling planet, possessions are important to you—sometimes too much so. Question your motivation if your home is cluttered. Emotional insecurity could be the reason you find it tough to let go. On the plus side, however, some of those things could turn out to be collectibles that you could sell.

Your Relationships

You're kind, caring, and ever sensitive to other's feelings, needs, and concerns. Friends know they can depend upon you, and you're always the first to lend a hand, as others do for you. This enhances a true friendship, but it also makes you prone to hard luck stories. Use your enviable common sense before others have a chance to take advantage of you and your kind heart. Then guide them in the right direction so they can do more for themselves.

You feel complete when you have someone to love, but you're also highly selective, which makes it tough to connect with a potential mate. Take a chance now and then because first impressions aren't always accurate. Then, resist the urge to overanalyze every word and to play the "what-if" game. Use your innate ability to set the perfect mood for romance and enjoy the moment. The depth of your feelings and your devotion know no bounds, and you expect commitment to be for a lifetime. If possessiveness or even jealousy are a challenge, remind yourself that a relationship built on trust and sharing is much stronger, and the way to fulfill your desire.

Your loyalty to family runs deep, even when relations are strained. Proud of your family, you have strong emotional ties with siblings, and keep in touch with your extended family. Children are a joy, although your tendency more often than not is to expect perfection. High standards are admirable and necessary, but expecting your children to excel in any every activity is unreasonable. Help them discover their individual skills and talents and then encourage them through constructive criticism. Be their No. 1 cheerleader.

Your Career and Money
You're attracted to a career that offers stability as well as one that offers a high level of autonomy. You also place a high value on the right mix of teamwork and individual effort. You can rise to the top in a flash, but your career life is also prone to sudden changes, so always keep your options open and resist the tendency to stay put if opportunity comes your way. A congenial, aesthetically pleasing workplace environment is a must. Without it you find it difficult to perform to the best of your ability. Communication is almost as important, and you enjoy one-on-one brainstorming and problem-solving, both of which can increase productivity. Although workplace conflict upsets you more than most, it's often better to settle disputes than to ignore them. Make an effort to acquire confrontation and conflict resolution skills if this is an issue for you.

You have a financial edge because Taurus is one of the universal money signs. This gives you a sixth sense in money matters, which, when combined with your practical nature, can help you build substantial net worth. Although you rarely splurge, you will treat yourself to luxury items—if the price is right. Good value matters more to you than the price tag, and your financial radar often guides you to the best bargains around.

Your Lighter Side
Your quest is comfort—a comfy chair of your own, a comfy bed, comfy clothing, and comfort food. But what many people don't realize about you is that your mind is almost always on the move even if your body isn't. You're quick to grasp what others miss, a trait that's useful for sizing up people and situations.

Affirmation for the Year
Change is personally empowering.

The Year Ahead for Taurus

Jupiter spends much of the year in Pisces, your solar Eleventh House of friendship, groups, and networking. The year 2010 is thus a "people year," and the more you connect with others, the greater your luck and opportunities. Friends will have a central role in your life, and you can widen your circle through a club, organization, or volunteer activity, or by organizing a community or neighborhood action group. So get motivated, get out of the house and your comfort zone, and start networking. Some of these people will become treasured friends, and others will benefit your career and personal aims during the next twelve years. The end of May and first few days of June are particularly auspicious because Jupiter, planet of luck, will merge with Uranus, planet of the unexpected. Someone you meet then could turn your life around, and the same could occur in mid-September when these planets meet again in Pisces.

The Eleventh House is closely aligned with fun and friendship, but it also has another, more obscure side: goals and objectives. This is the year to carefully define your hopes and wishes, what you want to achieve in the next twelve years. Dream big! Then use your practical mind to create and refine a short- and long-term plan to achieve what is realistically possible for you. But don't be quick to discount an idea just because you've never explored or tried it before. With your notable determination you can achieve more than most.

Jupiter makes a brief visit to Aries, your solar Twelfth House, June 5–September 7, to give you a preview of 2011. Placed here, Jupiter is often the ultimate lucky charm, the eleventh-hour stroke of good fortune just when you need it most. But you'll also have the urge to kick back and simply enjoy lazy summer days. Resist! With the changes indicated by this year's other planetary alignments, your best option is to use this time wisely. Strengthen your body, your mind, and your job prospects. By expending the effort you'll make your own luck, and thus maximize Jupiter's potential.

Saturn also visits two signs this year. But where Jupiter's foray into Aries gives you a glimpse of the future, Saturn's retreat into Virgo offers the opportunity to complete what was begun in 2007, when Saturn entered Virgo. Taurus and Virgo are both earth signs, so the energy is compatible, meaning you can get the best of Saturn in your solar Fifth House. Placed here, the emphasis is on children, creativity, and romance. Reflect on the related events that have occurred in the past few years and prepare yourself to wrap up unfinished business this summer when Saturn completes its Virgo tour.

The only wrinkle here is that Saturn has not been operating on its own since November 2008, when it first aligned with Uranus in Pisces across your solar Fifth/Eleventh House axis. Change (Uranus) has thus been competing with stability (Saturn), and challenging you to cling to the known while embracing the unknown. You could realize success this summer if, for example, you've nurtured a creative endeavor over the past few years with the goal of turning it into a home-based business. Or, one of your children could "suddenly" excel in academics or athletics when effort and hard work finally pay off. A romance that has been up and down, or possibly just drifting along, is likely to reach a turning point that results in commitment or moving on. The Fifth House also governs investments and games of chance, both of which require caution because of Uranus's unpredictability and Saturn's insistence on responsibility. Caution is also advisable with new people (or even old friends) who encourage you get involved in a "can't fail" venture; temper optimism with common sense. On another level, you might finally decide it's time for a career change. Not that this will be anything new because the energy has been building for several years; now, however, you'll be ready to move in a new direction.

Saturn begins and ends the year in Libra, your solar Sixth House of daily work, which reinforces the possibility of a career or job change. This may come about as a result of downsizing at your company, whether it's your job that's directly affected, coworkers' jobs, or an industry-wide trend. Developments are likely May 27–August 12, when Saturn in Libra will be directly opposite Uranus, which transits Aries, your solar Twelfth House. The Sixth and Twelfth Houses are also associated with wellness, so listen to your body. Schedule checkups, and be proactive if you could benefit from a healthier diet and regular exercise. As much as Saturn wants to hang onto the status quo, this is one way you can use Uranus to advantage to change what needs to be changed. New habits adopted this summer will yield positive results for years to come.

Uranus also travels in two signs during 2010. It spends most of the year in Pisces, your solar Eleventh House of groups and friendship. From May 26 to August 12, however, this planet of the unexpected will transit Aries, your solar Twelfth House of self-renewal. This makes summer a good choice for vacation time, when you'll be ready for a break. Do more. Look inward and think about personal changes that could help ease stress and calm your mind.

Neptune in Aquarius continues its long tour of your solar Tenth House. Although it forms no major planetary alignments this year, it's still active and can be influential in your career. Neptune inspires but also disappoints; nurtures creativity but also confuses. Tapping the best of this planet can be a challenge at times, although you're likely to come out ahead in the outer world if you question before you act and resist the urge to put too much faith in the status quo. What you're told may be true but idealistic, or illusion rather than reality.

Pluto in Capricorn, your solar Ninth House, brings education to the forefront, so you may explore the idea of returning to school to begin or complete a degree. The Ninth House also rules spirituality (and religion), legal affairs, and long-distance journeys; thus, any of these areas may be active. However, Pluto's main thrust is transformation—a profound change in thinking, beliefs, or life direction—so don't be surprised if you begin to question long-standing principles. Here, too, change can be a very good thing. Think long term.

What This Year's Eclipses Mean for You

The first of this year's eclipses (solar, January 15) occurs in Capricorn, where it's beautifully aligned with Venus in the same sign, Uranus in Aquarius, and your Sun sign. This means you could be traveling in the coming months on business, pleasure, or both. Relocation for a job opportunity is also a possibility, but this eclipse, like the year's second one, also in Capricorn, suggests further education as a link to career success. Tune in to the energy of the June 26 Capricorn lunar eclipse, which directly aligns with Pluto in Capricorn and Mercury in Cancer, your solar Third/Ninth House axis of learning. Just as significant is that this eclipse aligns with Jupiter and Uranus in Aries, which results in a multi-planet configuration that demands action. This eclipse also advises caution in financial matters involving a friend, partner, or romantic interest. Optimism could get the best of you and your wallet.

The July 11 solar eclipse in Cancer, your solar Third House, emphasizes the importance of communication. You're also likely to be on the go (more than you wish) to the point that your car might feel like a second home. Some of this could involve your children's busy schedules and regular trips to the library to fuel a growing passion for reading and learning. Check into what your city offers in leisure-time learning or sports classes. The emphasis shifts to finances with the December 21 lunar eclipse in Gemini (Sun in Sagittarius) that spans your solar Second/Eighth Houses. Although you could luck into a nice windfall because this eclipse is aligned with Jupiter and Uranus in Pisces, big risks could dent your net worth. Nevertheless, this is the time to take a chance on the lottery or a sweepstakes entry—but not to enter into a financial partnership no matter how much you trust the other person.

Saturn

If you were born between May 18 and 20, Saturn favorably contacts your Sun from Virgo, April 7–July 20. Placed in your solar Fifth House of romance, children, pleasure, and creativity, you can expect related activities to claim a larger share of your time. What's more, you'll benefit from Saturn's steadying influence and the stamina to follow through and complete unfinished projects. Find a practical outlet for your creative energy that yields tangible results, such as gardening, woodworking, or arts and crafts. You'll be able to exceed even your high expectations and get great personal satisfaction from the effort. The same is true of an exercise program or learning a new sport, or getting involved with your children's athletic activities as a coach or fund-raiser.

Saturn can help you rejuvenate and strengthen ties in a long-term relationship, or you might decide it's finally time to commit to lifetime togetherness with someone you've been seeing for several years. It's also possible you'll reconnect with a former love interest and rekindle the spark. Think first, however; remember the good times as well as why you parted company.

Saturn in Libra will contact your Sun **if you were born between April 19 and May 6**. However, its influence will be minor between mid-August and year's end, **if your birthday is after April 25**. Treat yourself well when you feel this low-energy period arrive, get plenty of sleep, and take precautions to avoid a cold or flu.

If your birthday falls in the April 19–25 time frame, you'll experience the effects of Saturn in Libra between January 1 and April 6 and from late July through mid-August. Job stress could rise during this period, making it all the more important to safeguard your health. A checkup is a good idea, and you should consult a medical professional if a health issue arises; otherwise, it could become a chronic problem, which Saturn is noted for. You also can use Saturn's influence to be proactive: fuel your body with healthy foods, replace bad habits with good ones, commit to regular exercise (begin slowly after a medical consultation), and strive for a balanced lifestyle that's the right mix of work and play.

Uranus

The odds favor change on your terms **if your birthday is between May 13 and 20**, mostly because Uranus contacts your Sun from Pisces, your solar Eleventh House. This transit may nudge you in new directions that could be stressful because Saturn will align with Uranus at the same time. You will need to take charge of matters concerning friends, groups, children, romance, leisure time, and investments. You'll be glad you did, because the universe could take action if you don't. Plan and make decisions about the relevant Fifth/Eleventh House people and activities the first part of the year, and don't hesitate to put you and your interests first. Then you'll be ready to resolve matters between April and July. If you're a parent, give your children extra attention and get involved (or more involved) in their lives. They can benefit from your wisdom as much as you can from theirs.

Uranus in Aries will contact your Sun between May 27 and August 12 **if your birthday is April 19 or 20**. Job stress could rise now, so safeguard your health and do whatever works for you to ease tension. Exercise and hobbies are more productive than worry and better for your overall health. Also take some time to listen to your inner voice. You'll feel the initial rumblings of needed, desired, or even impending change begin to surface.

Neptune

If you were born between May 14 and 20, you'll experience the many facets of Neptune in Aquarius, your solar Tenth House, as it contacts your Sun. This planet of illusion has the potential to initiate instant fame (or infamy), or simply make you wish you could be the next big-screen star. Try to find the middle ground to get the best of Neptune. Do that and you'll benefit from the almost magical aura of popularity this mystical planet can bestow. On the flip side, however, is a tendency for irresponsibility (yes, even for you!) and a belief that rules can be broken—without consequences. Curb that feeling. Find a trusted friend or mentor who can confirm or challenge your career decisions. Even though you believe your thinking is on track, it may not be.

Pluto
Pluto's influence is far-reaching, whether you experience it on a personal level, a global one, or both. You can thus anticipate major change this year **if your birthday is between April 22 and 24**. Placed in Capricorn, your solar Ninth House, Pluto can trigger legal matters, education issues, travel difficulties, in-law challenges, relocation, or a spiritual awakening. Even if you don't experience these directly, at least some of them will affect those around you in your personal or work life. It's important to remember that Pluto triggers needed—not unnecessary—change as a means of removing the old to make way for the new. True, this transformational process can be difficult and stressful, but it almost always has an ultimately favorable outcome. And because Pluto forms a beneficial contact with your Sun you can manage this Pluto transit with success.

Saturn-Jupiter-Uranus-Pluto
If you were born in the April 19–23 time frame, you'll experience the full influence of this major four-planet lineup involving your solar Sixth, Twelfth, and Ninth Houses. This is a challenging lineup at best, and as a result you could find yourself searching for a job or taking on increased responsibilities because of company downsizing. Be alert and aware of what's going on in the world, in your industry, and with your employer. Although Jupiter will offer some protection from its solar Twelfth House position, this expansive planet can also make a mess even messier. On the upside, however, this might be just the motivation you need to return to school to complete a degree, switch careers, or to get advanced or specialized training. Consider this even if your job is stable.

Taurus/January

Planetary Hotspots
The domestic scene could be tense at month's end when the Full Moon joins forces with retrograde Mars in Leo, your solar Fourth House. A repair may be needed (leave it to a pro) and family members could be at odds. Keep your cool and keep your pointed comments to yourself. Also be sure candles are out and the stove off before you leave or head for bed.

Planetary Lightspots
The January 15 New Moon (solar eclipse) in Capricorn is great incentive for a winter getaway. Make reservations for a week or weekend with family or friends, or make it a romantic rendezvous designed for two. Limited time? Try a local hotel or spa.

Relationships
January 17 marks the start of your year of friendship and socializing. That's when Jupiter enters Pisces, your solar Eleventh House. Networking can multiply your luck, and you could expand your circle by getting involved in a club or organization. Just be careful not to take on too much because optimism can get the best of you.

Money and Success
You're among the favored few when Venus enters Aquarius, your solar Tenth House, on the 18th, followed by the Sun. That's a real plus for your career and status. But there's another planetary force at work. Saturn in Libra, your solar Sixth House of daily work, clashes with Pluto in Capricorn, your knowledge sector. Think seriously about additional schooling as a way to increase your marketability and earnings.

Rewarding Days
1, 4, 7, 8, 12, 14, 17, 18, 19, 24

Challenging Days
3, 6, 13, 16, 20, 21, 23, 27, 30

Taurus/February

Planetary Hotspots

Life is mostly status quo this month, perking along under favorable planetary alignments. The exception is Mars, which is still retrograde in Leo, your solar Fourth House. That and February's career emphasis can cause conflict on the home front. Strive for balance to get the best of both by giving loved ones your full attention at home.

Planetary Lightspots

Add some fun to your life! Play! The February 28 Full Moon in Virgo highlights your solar Fifth House of creativity, romance, children, and leisure-time activities. This is a terrific influence for singles looking for a match, as well as couples in love. Relax evenings and weekends with your kids and spend some time with a favorite hobby. Or get motivated and stretch your muscles at the gym.

Relationships

You'll have plenty of opportunities to get together with friends this month, thanks to Venus in Pisces, your solar Eleventh House, from the 11th on. Get a group together for an evening out, plan a day trip, and accept invitations that come your way. If you're single you might want to take a second look at a pal. A romance could bloom with someone you've always considered a friend.

Money and Success

Your star continues to rise this month under the influence of the February 13 New Moon in Aquarius, your solar Tenth House. Career matters also get a boost from Mercury in Aquarius after the 9th, when you and your ideas will get plenty of attention. Listen to your intuition and take note of your dreams, both of which can be an asset to your worldly success.

Rewarding Days

1, 4, 6, 10, 13, 14, 15, 16, 24

Challenging Days

2, 5, 9, 17, 19, 20, 21, 23

Taurus/March

Planetary Hotspots
With several planets transiting your solar Twelfth House this month you'll want to be sure to get plenty of sleep and rest. That will help prevent a cold or flu, to which you're especially prone now because of a heavier workload. Promise yourself at least thirty minutes each evening to unwind and enjoy quiet time with yourself, a loved one, or a best seller.

Planetary Lightspots
Domestic tension eases after Mars turns direct in Leo, your solar Fourth House, on the 10th. That's the first item of good news. The second is you can now put this planet to work for you around the house. Get a jump start on spring cleaning and move on to routine repairs. If you feel especially ambitious, consider redecorating a bedroom or bathroom in spring colors.

Relationships
Your social life continues to be in high gear, thanks to the March 15 New Moon in Pisces, your solar Eleventh House. You'll especially enjoy quiet evenings with your best friends. But also join coworkers and acquaintances whenever you have the chance because this is a terrific month for networking.

Money and Success
The work pace accelerates under the March 29 Full Moon in Libra, your solar Sixth House, an influence that continues into the first two weeks of April. That's a great match with the current networking emphasis, which could open up possibilities for a new position. Keep it quiet, though, if you decide to pursue another job. Someone you trust could use the knowledge for personal gain.

Rewarding Days
4, 5, 9, 10, 12, 13, 14, 15, 19, 27, 28

Challenging Days
1, 2, 8, 11, 16, 20, 23, 25, 29

Taurus/April

Planetary Hotspots
Saturn retreats into Virgo, your solar Fifth House, forming an exact alignment with Uranus in Pisces on the 26th. A dating relationship could end under this planetary influence because either one or both of you are unwilling to adapt or commit. If you're a parent, get to know your children's friends, especially if you have teens. Peer pressure could overcome common sense.

Planetary Lightspots
Turn on the charm and wow everyone you meet. With Venus in your sign through the 24th your charisma will be at its best, as will your powers of attraction. Share your wishes with the universe and fully expect Venus to deliver, whether you're in search of love, money, success, or another desire.

Relationships
The only wrinkle in an otherwise great month for relationships is Mercury. It turns retrograde on the 17th in your sign. Think before you speak. Even someone with your top-notch people skills could trigger a misunderstanding now. Also be sure to check dates, times, and places before heading to an appointment or social event.

Money and Success
Venus moves on to Gemini, your solar Second House, on the 25th. This positive financial influence will be with you through mid-May, and could be just what's needed to earn a nice raise. It's also a plus for bargain shopping. Browse sale racks for career clothes and household items, as well as gifts for upcoming birthdays and even the holidays.

Rewarding Days
1, 5, 6, 8, 10, 11, 16, 20, 21, 27, 29

Challenging Days
2, 4, 7, 12, 19, 22, 23, 24, 26, 28

Taurus/May

Planetary Hotspots
You'll feel the push-pull of the past and the future as Saturn in Virgo aligns with Jupiter in Pisces across your solar Fifth/Eleventh House axis. Someone from the past, a former love interest or friend, could reappear, but the reunion will be bittersweet with little potential for the future. Or someone in your current circle could disappoint you by failing to keep a promise. Be optimistic yet also skeptical, especially if someone seeks your money.

Planetary Lightspots
Life is all about you! The May 13 Taurus New Moon has you in the spotlight of a new solar year with limitless possibilities. Make a point this year to express your individuality through a creative project. Who knows? It might eventually become a sideline moneymaker.

Relationships
Relationships are positive for the most part this month, especially after Mercury resumes direct motion on the 11th. But the best comes on the 19th when Venus enters Cancer, your solar Third House of communication. You'll have an ultra-charming way with words—at least with most people. The exception is controlling people, one of whom will push your limits. It may be a relative or a disagreeable neighbor. Tune out and keep your cool.

Money and Success
The May 27 Full Moon in Sagittarius shines brightly in your solar Eighth House of joint resources. That has all the potential to boost family income. But keep the information to yourself and curb your generosity. Take care of you and yours first before donating to a good cause, no matter how much you believe in it. Opt instead for a larger donation of your time and effort.

Rewarding Days
3, 7, 8, 12, 17, 21, 30, 31

Challenging Days
2, 4, 9, 16, 20, 22, 23, 26, 29

Taurus/June

Planetary Hotspots
The June 26 Full Moon (lunar eclipse) in Capricorn might trigger the desire for travel, or the need to care for an out-of-town relative. But a pleasure trip isn't the best idea because planetary alignments increase the odds for delays and cancellations. Also be cautious on the road and catch a ride if you plan to be out and about socializing. Conflict with difficult people or an authority figure is possible.

Planetary Lightspots
Take note of your dreams this month, especially during the first week of June when Jupiter and Uranus join forces in Aries, your solar Twelfth House. They can be especially insightful regarding your career and money matters. Listen to a hunch. It could guide you to the right place at the right time to cash in on a terrific opportunity.

Relationships
Mars dashes into Virgo, your solar Fifth House, on the 7th. That's a real plus for summer fun and outdoor events, as well as getting involved in your children's sports activities. With Venus in Leo, your solar Fourth House, from the 14th on, family relationships will be upbeat. Romance zings for couples in love, and singles could meet a sensational someone this month or next.

Money and Success
Money matters will be in focus under the June 12 New Moon in Gemini, your solar Second House. The lunar energy is promising for increased income, but it may be matched by increased expenses. You'll also have the urge to splurge midmonth, so plan ahead to shop only for necessities. But the same time frame could trigger a lucky find or a small windfall.

Rewarding Days
4, 6, 8, 9, 13, 14, 17, 18, 22, 27

Challenging Days
2, 3, 5, 19, 24, 25, 26, 29

Taurus/July

Planetary Hotspots

Saturn returns to Libra on the 21st and forms an exact alignment with Uranus in Aries on the 26th across your solar Sixth/Twelfth House axis. Job-related stress and strain accompanies this lineup when change occurs in your work environment. Downsizing is possible, and that may require you take on a heavier load. This is also your wellness sector, so treat yourself well, get a checkup if it's been a while, and find an outlet to release tension on a daily basis.

Planetary Lightspots

Take off for the weekend when the July 11 New Moon (solar eclipse) in Cancer, your solar Third House, encourages you to take a break. If that's unrealistic, relax with the full treatment at a day spa or spend the night at a local hotel just to get away from it all.

Relationships

Family relationships are in an upbeat period all month, and you'll delight in spending time at home with loved ones. What's more, with the right incentive (bribe!) they'll lend a hand with household projects. You might even be willing to part with a few possessions, which could net cash at a yard sale or consignment shop.

Money and Success

Although your job situation will be stressful, you're well-placed to take advantage of career opportunities, thanks to the July 25 Full Moon in Aquarius, your solar Tenth House. A new position could come your way, or you might be offered a promotion. Think carefully before you leap, however, because you may not have all the facts. The same applies to a home-based business. Although the idea might have promise, it will be tough to make a go of it. Job hunting? Send out résumés under the New Moon.

Rewarding Days
1, 6, 7, 11, 14, 15, 20, 24, 28, 29

Challenging Days
3, 9, 10, 12, 16, 21, 22, 23, 25, 30

Taurus/August

Planetary Hotspots
Your work life remains unsettled as Jupiter and Pluto add their energy to last month's Saturn/Uranus alignment. Downsizing is again possible, as is time away from work to care for a relative. Avoid legal matters if you have a choice, and be sure to document any job difficulties or significant activities just in case someone takes issue with your actions. Again, this planetary influence can bring health and wellness issues into focus, so be good to yourself.

Planetary Lightspots
Leisure-time activities are a great alternative to this month's stress. With Venus in Virgo, your solar Fifth House, through the 5th, and the Sun in the same sign later this month, you'll have plenty of opportunities to socialize and enjoy your favorite activities. Make them a priority evenings and weekends.

Relationships
The August 24 Full Moon in Pisces, your solar Eleventh House, highlights your friendship sector. Make a point to get together with your favorite pals, and also set the networking gears in motion so you can take full advantage of this avenue in September. It may take until then for people to follow through because Mercury turns retrograde in Virgo on the 20th. If you're undecided about a dating relationship, wait until late next month when you'll have a clearer picture.

Money and Success
This month's hotspot could be the initiative you need to complete your education or to prepare for a new career field. Ask someone close to you for advice and then act on it to open up new paths to the future. Also consider a short-term course of study to get up to speed on the latest technology in your career field.

Rewarding Days
2, 3, 7, 11, 15, 21, 26, 29, 30

Challenging Days
4, 5, 6, 10, 12, 16, 19, 20, 27

Taurus/September

Planetary Hotspots
The hotspot planetary energy of the past few months continues, but also begins to wane. And now you'll have a lucky charm on your side in the form of a Jupiter-Uranus merger in Pisces, your solar Eleventh House. Friends and networking contacts are your path to good fortune, and one of them could be your connection to a terrific opportunity. Focus your dreams on the future and refuse to be deterred if someone tries to derail your plans.

Planetary Lightspots
Summer fun peaks at the September 8 New Moon in Virgo, and with Mercury in the same sign resuming direct motion on the 12th, you're perfectly placed to fill your calendar with social events. The more people you meet, the greater your networking opportunities. The lunar energy triggers a new romance for some singles that could lead to commitment within a few years.

Relationships
Your solar Seventh House of relationships comes alive when Venus enters Scorpio on the 8th, followed by energetic Mars on the 14th. You'll be drawn to people and they to you, and you can gain all the support you need from close friends and family. This is a terrific time for couples, and for connecting with people at a distance.

Money and Success
Although workplace tension begins to ease, you'll still have some tense moments when people try to block progress. Put your determination to good use, both to jump hurdles and to complete every project on time and to the best of your ability. With persistence you can rise above the rest and overcome detractors. Fuel your effort with healthy food and plenty of sleep.

Rewarding Days
2, 8, 10, 11, 12, 16, 17, 20, 22, 27

Challenging Days
1, 3, 4, 9, 15, 18, 23, 25, 26, 30

Taurus/October

Planetary Hotspots
Venus, your ruling planet, turns retrograde on the 8th. Prepare yourself to be frustrated when personal plans and projects stall. Think of this six-week period as a time to regroup. Review the year's progress and set goals for the next six months. This is not the time for romantic commitment or to establish a professional partnership. Hold off until late November or December. By then you could have a change of heart and mind.

Planetary Lightspots
The October 22 Full Moon in Aries encourages you to step out of your hectic world and slow the pace a little. You'll enjoy time alone, and relaxing evenings at home with your favorite people far more than attending social events. Catch up on sleep and reading, watch a few movies, and treat yourself to some comfort food.

Relationships
Relationships will be mostly easygoing, especially after Mercury enters Scorpio on the 20th, followed by the Sun a few days later. With Mars in the same sign through the 27th, you'll delight in lively conversation with friends and family, and could gain fresh insights into people and their personalities. Ask for favors, which are likely to be granted, and do the same for others.

Money and Success
This month's New Moon on the 7th in Libra, your solar Sixth House, infuses your work life with positive, upbeat energy. Not that you won't have challenges. You will. But successes will outnumber setbacks despite the hard work you have in front of you. Do some subtle self-promotion by sharing your accomplishments with decision-makers.

Rewarding Days
1, 7, 8, 9, 10, 18, 19, 24

Challenging Days
3, 6, 13, 16, 17, 20, 27, 31

Taurus/November

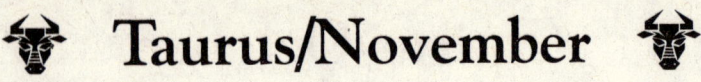

Planetary Hotspots
The November 21 Full Moon in Taurus is both a lightspot and a hotspot. You're in the lunar spotlight, ready to stride toward achieving your desires, goals, and wishes. The challenge, however, is that this Full Moon clashes with Neptune in Aquarius, your solar Tenth House, so you may find it tough to decide what you want, especially in your career and the outer world. Think first. Then, take action.

Planetary Lightspots
Circle November 18 on your calendar. That's the date Venus, your ruling planet, resumes direct motion and your life begins to regain momentum. Placed in Libra, your solar Sixth House of daily work, Venus's direction change complements this month's Full Moon influence in your career. Seek people who can help advance your aims.

Relationships
November is one of the best months of the year for relationships, thanks to the November 6 New Moon in Scorpio, your solar Seventh House. Positive planetary alignments enhance the effect, and you'll be among the most popular on the holiday social scene. To take advantage of this potentially lucky time, get together with friends, meet new people, and continue to pursue networking opportunities.

Money and Success
Mars in Sagittarius, your solar Eighth House, could trigger a raise or bonus. More to the point, you'll have the energy and initiative to do all you can to boost your bank account. Mercury advances in Sagittarius from the 8th to the 29th. Use it to organize financial records, learn more about money management, and make budget decisions for 2011.

Rewarding Days
1, 2, 5, 6, 14, 16, 19, 20, 28

Challenging Days
3, 9, 15, 17, 18, 23, 24, 27, 30

Taurus/December

Planetary Hotspots

Mercury turns retrograde in Capricorn, your solar Ninth House, which can complicate travel and long-distance communication. But there's more. Mercury retreats into Sagittarius, your solar Eighth House, on the 18th, before resuming direct motion on the 30th. And that can affect finances. Carefully check your credit card statements, keep gift receipts, and pay bills early. Be cautious if you order online; mix-ups could easily occur with billing and shipping charges.

Planetary Lightspots

Holiday travel is appealing with Mercury, Mars, and the Sun transiting Capricorn. You could take your chances with retrograde Mercury or opt for an alternative. Encourage friends or family to do the traveling and welcome them to your home. Chances are, you'll hear from many out-of-town friends this season.

Relationships

Venus advances in Scorpio, your solar Seventh House, all month, making favorable connections with many other planets. That's a terrific influence for laid-back evenings and lively weekends with your partner, close friends, and family. Stop the clock and enjoy every moment.

Money and Success

This month's New Moon (December 5) in Sagittarius and Full Moon/lunar eclipse (December 21) in Gemini highlight your financial sectors. Although retrograde Mercury can interfere, money matters are generally in positive territory and potential is high for you or your mate (or both of you) to receive a nice year-end bonus. Treat yourself to something special, but don't go all-out; save instead. Friends and coworkers may surprise you with gifts, so be prepared, and remember that homemade goodies are ideal for some.

Rewarding Days

2, 3, 8, 10, 12, 13, 29, 30, 31

Challenging Days

7, 9, 14, 15, 20, 21, 26, 27, 28

Taurus Action Table

These dates reflect the best—but not the only—times for success and ease in these activities, according to your Sun sign.

	JAN	FEB	MAR	APR	MAY	JUN	JUL	AUG	SEP	OCT	NOV	DEC
Move			25, 26			15, 16	12, 13	9, 10				
Start a class		19–28	1–15		17	25–30	1–8	7	3	1, 28, 29		
Join a club		19–28										
Ask for a raise	25		22		14	11				4–7		
Look for work						19, 20		13		4–19		
Get pro advice	8		4, 5								1–5	2
Get a loan						24					26, 27	
See a doctor	21, 22		16, 17, 25, 26					13				
Start a diet	6, 7	2–4	2, 3, 29								3, 4, 30	1
End relationship	8, 9	5, 6	4, 5, 31	1, 2, 29								3, 4
Buy clothes	4, 5, 31				21, 22		15, 16, 28–31	11, 12	11, 12			
Get a makeover		19, 20		5, 6, 10, 11, 15, 16	12, 13		3					
New romance				15, 16								
Vacation	1–15											22–31

GEMINI

The Twins
May 20 to June 21

♊

Element: Air
Quality: Mutable
Polarity: Yang/masculine
Planetary Ruler: Mercury
Meditation: I explore my inner worlds
Gemstone: Tourmaline
Power Stones: Ametrine, citrine, emerald, spectrolite, agate
Key Phrase: I think
Glyph: Pillars of duality, the Twins

Anatomy: Shoulders, arms, hands, lungs, nervous system
Color: Bright colors, orange, yellow, magenta
Animal: Monkeys, talking birds, flying insects
Myths/Legends: Peter Pan, Castor and Pollux
House: Third
Opposite Sign: Sagittarius
Flower: Lily of the valley
Key Word: Versatility

Your Strengths and Challenges

Gemini is the communicator of the zodiac, and you're a master at both tracking down information and collecting a vast array of trivia, all of which is stored in your clever, Mercury-ruled mind. But communication is designed to be a two-way exchange, something some Geminis forget. Plus, the more you listen (and keep information to yourself), the more you learn.

You're also the multitasker of the zodiac, the one who can almost effortlessly juggle the hectic demands of everyday life. Doing two or even three things at once is a snap for you. With so much going on, however, organization is vital. To-do lists, a calendar, and a PDA or planner are essential.

Witty and charming, your lighthearted personality boosts your popularity in business and social situations. You're also a natural networker who brings people, ideas, and plans together. Your tendency to think out loud, however, can be misunderstood. So choose your words with care, because not everyone realizes some of what you say is brainstorming rather than firm opinion. Geminis have short attention spans. It's thus difficult for you to stick to one task for a long stretch. The exception is anything that really captures your interest. Then your powers of concentration are at their best as you satisfy your curiosity. In most situations, however, a unique approach can help maintain interest; switching tasks is also helpful as long as you achieve the ultimate goal—completion.

Your Relationships

You enjoy meeting people and welcome new friends, although all but a select few are more in the acquaintance category. Friends also have a tendency to appear and disappear from your life—here today, gone tomorrow, almost as if they arrived to fulfill a specific purpose. This keeps your life as lively as you like it, but it can make it tough to develop a wide circle of people you can depend on.

Flirtatious and witty, you want relationships with people who cherish freedom and independence as much as you do. Long talks are one way to your heart and intellectual rapport is a must before you'll even consider a serious dating relationship. But it can be tough for you to express your feelings because you're so focused on thoughts. That's okay with casual romances but lasting ties require much more.

In love, you delight in togetherness—if you can also pursue your own interests. You have much in common with Sagittarius, your opposite sign, and with the other air signs, Libra and Aquarius, or another Gemini. Leo and Aries also offer possibilities, but you may clash with a Virgo or Pisces. Your home is either a pristine showplace or a disorganized mess. There are no in-betweens with Gemini's domestic life. Family relationships are similar—formal and somewhat distant or active and engaging. Some Geminis grow up in a highly critical environment where perfection is the only acceptable norm, yet nearly impossible to achieve. Remember this if you're a parent and resist the natural urge to do the same.

Your Career and Money
Career success and happiness are linked to the freedom to explore new ideas and to use your creative energy. A loosely structured environment is best, and you excel when given challenging tasks. Unfortunately, you can be drawn into power plays at work, primarily because of your idealistic career viewpoint. Balance optimism with reality and don't expect others to always play by the rules. Be aware of what's going on around you even when you'd rather be immersed in a task. Your financial success depends almost entirely on your attitudes and habits regarding spending, saving, and investing. Train yourself to think long-term and to remove emotions from financial decisions. If you spend to feel good, find a substitute that will deliver the same reward without the expense. For many Geminis, income rises in later years, after age forty, and with careful planning you can be in a position to enjoy a comfortable retirement.

Your Lighter Side
Your mind is one of your greatest assets, and it rarely slows as thoughts and ideas zip along at lightning speed. Keeping track of them is your challenge, but one you can certainly meet. Keep a notebook handy to write things down so you don't forget all those great bits of wisdom. Then use it for reference to boost your already excellent problem-solving and decision-making skills.

Affirmation for the Year
I welcome the opportunity to achieve my career best and to create financial security.

The Year Ahead for Gemini

Jupiter spotlights your career sector as it travels in Pisces, your solar Tenth House, much of the year. This transit can be your lucky charm, especially later this year when Jupiter merges with Uranus in Pisces. Resist the urge to slide along now. Otherwise, you won't reap all the potential benefits Jupiter can bring. Network to the max and cultivate those who can further your ambitions. But don't take everyone at face value because some will promise the world and fail to deliver. You'll also want to think carefully if a home-based business sounds appealing. Although you might be able to make a go of it, this year's planetary alignments make that chancy. Opt to start slow, no matter what anyone urges you to do.

You get a preview of 2011, when Jupiter briefly visits Aries, your solar Eleventh House, June 5 through September 7, and gives your social life a boost. June may bring a chance encounter with someone who could open doors, but be cautious if you hear grand promises. They may or may not come to fruition. If you're involved in any group activity, you could find yourself catapulted into a leadership position, which could further your ambitions. However, there may be more time and effort required than you bargained for, so ask questions and take on only what you can realistically handle.

Saturn begins and ends the year in Libra, with a brief (and final) retreat into Virgo, your solar Fourth House, April 7–July 20. Saturn in Virgo will help you to conclude domestic matters begun at any time in the past three years. Relocation is possible, but a home purchase or remodel is best postponed until at least this fall. In any case, it would be a mistake to overextend yourself based on hoped-for or promised career advancement. You also may find yourself caring for an elderly relative, or an adult child either leaving or returning home.

This is, however, a positive time to get your home in shape, to de-clutter, and take care of do-it-yourself household repairs. If you've been trying to sell your home, Saturn's final transit could bring success. But it's also possible you will net less than the asking price.

Saturn is in Aries much of the year, spotlighting your solar Fifth House of romance, recreation, children, and creativity. Despite what you may have heard, Saturn here does not mean your love life will grind to a halt. Far from it! More likely, you'll take matters of the heart more seriously, and if you're single you could rekindle a relationship, enter into a long-term relationship or one in which there is a large age difference. For couples, this is a time to take love seriously, or to rediscover each other. You may get more involved in your children's lives and take parenting much more seriously than in the past. That's positive if you set limits on yourself; otherwise, your children could resent what they perceive as an attempt to control their lives. Teach them to be responsible for their actions and give them a reasonable amount of freedom to learn from their mistakes. You might also decide the time has arrived for an addition to the family, or learn that a new arrival is unexpectedly on the way.

This Saturn placement is excellent for creative endeavors and hobbies that produce practical results. Give gardening or home decorating a try, or follow through on learning a leisure-time activity that's always been of interest. You might even eventually net a profit from this by turning a hobby into a sideline income. Be cautious and conservative, however, with investments. Even so, you may see the value of your 401K or portfolio drop as Saturn transits Libra the next few years.

You may have already experienced career changes since 2003. With Uranus, in your solar Tenth House, the unexpected is to be expected. But because of this planet's retrograde pattern, it's possible you have sensed coming changes. That might be an internal process, where you're warming up to the idea, or it could be happening around you, in your company or career field. Instability is the norm rather than the exception when Uranus is involved. One major plus this year is Jupiter, which merges its lucky energy with Uranus in Pisces. That could trigger the job of your dreams or put you in a position to explore new possibilities. Either way, be aware and keep your options open. But Saturn in Virgo also aligns with Uranus in Pisces. This challenging contact could point out the need to move on career-wise, again through your choice or your company's. If so, networking will be your most valuable resource. Think twice if you want to start your own business. General economic conditions caution against this idea and planetary alignments with your Sun sign make it risky at best.

Uranus briefly switches signs this year, traveling in Aries May 27–August 12. Use the time to expand Eleventh House activities such as networking, friendship, and clubs and organization activities.

Involvement in a professional group or a volunteer activity could open doors in 2011 by putting you in touch with new people. Your circle of friendship will expand during the next seven years. Some of them will arrive for a purpose and then disappear just as quickly, and others will become lifelong pals. What you'll gain most from this transit is an extra measure of luck—being in the right place at the right time to snap up opportunities.

Neptune continues its long trek through Aquarius, your solar Ninth House. Here, it emphasizes spirituality and knowledge, and encourages you to look beyond the obvious into the metaphysical. Faith is also a strong theme, although putting your trust in just anyone would be a mistake because some people will be quick to take advantage of your sympathetic nature at this time. This is an excellent influence if you have yet to complete your education or want to get an advanced degree or training. But you might have to push yourself even if you know it's in your best interests to broaden your knowledge. Neptune is great for dreaming, but less than effective for action. Yet with this year's other planetary alignments you can find the motivation within. So no matter where you are in life, give yourself the edge by expanding your knowledge base. Learning can benefit you now even if it's a leisure-time class.

Pluto, now firmly in Capricorn, your solar Eighth House, can multiply your wealth in the coming years or trigger sizeable losses. It's up to you to choose wisely regarding investments and other financial matters. Risk-taking is definitely unwise and you should be cautious about mixing money with friendship. However, you probably can be successful if you think long-term and let your money grow. With Pluto here you'll also have opportunities to travel in the coming years, both for business and pleasure.

Your earning power is likely to increase in the coming years. But rather than spend all that you earn (and maybe more), get in the habit of setting aside a big percentage for the future. Payroll deduction is an easy way to accomplish this. You'll never miss what you don't see, and a rising bank balance will provide the financial security you want and need.

What This Year's Eclipses Mean for You

There are four eclipses this year, two solar and two lunar. Three of them emphasize the planetary alignments and the fourth invites you to explore new personal territory. Each is active for approximately six months.

The January 15 solar eclipse in Capricorn spotlights your solar Eighth House of joint resources, as does the June 26 lunar eclipse. The January eclipse is beautifully aligned with Venus, also in Capricorn, and Uranus in Pisces, so it could trigger a minor windfall or pave the way to future opportunities. This is a favorable influence if you need a scholarship, for example, or want to take the trip of a lifetime. Save some, though, so you have resources to tap when necessary.

The June lunar eclipse in Capricorn may not be nearly so favorable, however, because it activates Pluto in the same sign and clashes with Jupiter and Uranus in Aries. There is no question that this lunar energy could fulfill your financial wishes, but your thinking may be more optimistic than realistic. So plan ahead to safeguard your assets, especially investments, and be skeptical if a friend proposes a financial endeavor or seeks a loan. Either one is likely to backfire despite your pal's promises.

July 11 brings the year's second solar eclipse, which occurs in Cancer, your solar Second House of personal resources. That's promising for financial gain, particularly in property deals. Begin house-hunting then if you're in the market for a new home and can do it without overextending yourself. This is also a promising time to find a rental, or to sell property. Keep in mind, though, that this eclipse may not trigger an overall gain, but simply provide a balance for the potentially negative effect of the June eclipse. Be cautious in money matters. This is terrific energy, however, to rid yourself of clutter and to find a place for everything. With luck, you could net some extra cash by holding a yard sale or taking usable items to a consignment shop.

You're in the spotlight at year's end with the December 21 lunar eclipse in your sign. This one favorably contacts Neptune in Aquarius, your solar Ninth House, but clashes with Jupiter and Uranus in Pisces, your solar Tenth House. Put the combination to work for you and step up to a career challenge that's also an opportunity. Just be sure you're fully informed about both the pluses and the minuses. Ask tough questions and accept nothing less than honesty. That's your best protection against potential misunderstandings that can emerge in the future. If possible, get things in writing.

Saturn

If you were born between June 19 and 21, Saturn in Virgo will contact your Sun between April 7 and July 20. With Saturn in your solar Fourth House, you can expect domestic and family matters to take priority during that time frame. Reflect on what occurred toward the end of 2009, and take the initiative to bring these matters to conclusion this spring.

You'll undoubtedly feel the pressure of increased responsibilities at home and/or for loved ones, which can lead to reduced stamina. Make rest and sleep priorities by sharing the load with another family member. You'll also tend to take life far more seriously than usual. One of the best antidotes for this and a heavy load is to socialize with friends—even if time alone is more appealing. Other people, especially close friends, can help you keep things in perspective.

Saturn will contact your Sun from Libra, your solar Fifth House, **if you were born between May 20 and June 6**. **If your birthday is May 26 or later**, however, you'll barely notice this transit because it will be active only a week or two between mid-August and year's end. **Geminis born between May 20 and 25** will experience Saturn's influence from January 1 through April 6 and again from July 21 through early August. Saturn's easy alignment with your Sun from the Fifth House encourages you to explore your creative energy, possibly with a partner, or by turning a hobby into a money-making venture. This year's planetary alignments do not favor that, however; it's wise to wait a while. The Fifth House is also associated with recreation and children, so you might enjoy getting involved in your children's sports or other after-school activities. You can have a major influence on them now, helping them to learn good study habits and to accept age-appropriate responsibilities. More important, listen to what they say because ultimately you will learn far more from them than they do from you.

You'll get the best of Saturn stamina because this planet contacts your Sun from a fellow air sign. It also will be easier to take charge of your personal life and to set goals and achieve them. For example, this is an excellent influence if you want to get in shape or step up to an overall healthier lifestyle. Not that it won't be work. It will be, but you'll also feel and see the rewards that come from Saturnian effort and appreciate them all the more.

Uranus

Career-related change is this year's theme **if your birthday is between June 13 and 21**, and even more so **if you were born between June 16 and 21**. You'll want to keep your options open and use your sixth sense to pick up on any developing trends at work. With the ultimate planet of unpredictability involved, this transit could just as easily trigger a fabulous opportunity as an involuntary job loss. Or there may be significant changes in your workplace that motivate you to look elsewhere. The most likely time frame for any of these scenarios is spring, when Saturn and Uranus line up across your solar Fourth/Tenth House axis. On the upside, Jupiter meets Uranus in Pisces this fall, giving you a window of opportunity for career gains. Keep that in mind in the preceding months so you're ready to take advantage of this good-luck charm.

It's important to realize that change is tough for everyone (even easygoing Gemini!). Change brings new endeavors, though, and also opens your mind to new ideas and personal directions. Don't be surprised if your usual restless energy increases and you begin to search for new horizons as the year unfolds. But let things jell rather than jumping at the first opportunity that comes along (or the first idea that occurs to you). It might or might not be the right move for you, but all will become clear this fall or early next year.

If you were born between May 20 and 21, you're among the first of your sign to have Uranus link its energies with your Sun from Aries, your solar Eleventh House. With the spotlight on friendship and group activities from May 27 to August 12, you have the opportunity to make new acquaintances, some of whom will be valuable networking contacts. Any sort of group endeavor will be more appealing. The more people you connect with the better. Some of them will enlighten you and others will spark fresh ideas for future achievements. It's also possible a chance encounter in June could be just the motivation you need to take the first important steps toward realizing a personal ambition put on hold many years ago.

Neptune

Neptune will contact your Sun from Aquarius, your solar Ninth House, **if you were born between June 15 and 21**. This a definite plus for creativity and intuition. So you'll want to find an appropriate outlet to express this side of yourself. Many Geminis enjoy writing; give this a try if you're among them. Take a creative writing class if becoming a published author is your goal. Or record your thoughts in a daily journal, which is a great way to deepen your self-understanding. Intuition is also associated with Neptune, so you'll want to listen to your inner voice this year. With practice you can strengthen this side of yourself. Calm your mind and listen. Then follow the promptings, testing yourself first on small things such as which route to take to work or where to shop for the best bargains. The more you do this the stronger your intuition will become. This transit does come with a potential downside, however. You'll be more trusting and concerned about people now, which makes it easier for others to take advantage of you. Question motives and look out for yourself.

Pluto

If you were born between May 23 and 26, Pluto will contact your Sun from Capricorn, your solar Eighth House of joint resources. The Eighth House encompasses everything from inheritance to insurance to your partner's income and creditors, so you can expect at least some of these matters to be in the forefront. Because this transit indicates a certain level of stress and strain involving finances, you'll want to spend less rather than more earlier in the year and make savings a priority. Be very cautious with investments and avoid loans if at all possible. This is the year to focus on becoming debt-free, or at least to make major strides in that direction. It's also possible you may need to forgo a raise or even experience a salary reduction as an alternative to downsizing by your employer.

Jupiter-Saturn-Uranus-Pluto

This unusual four-planet lineup will contact your Sun this summer **if your birthday is between May 20 and 24.** Jupiter and Uranus in Aries and Saturn in Libra will connect across your solar Eleventh/Fifth House axis, and all three planets will clash with Pluto in Capricorn, your solar Eighth House. An exciting romantic opportunity, or one that involves a friend or group, could come your way at that time. Look closely before proceeding, especially if it involves money. What appears to be a sure thing probably won't be, despite assurances and promises. Protect your resources. It's also possible one of your children may need your financial help.

Gemini/January

Planetary Hotspots
Between Mars retrograde in Leo, your solar Third House, and the Full Moon in the same sign on the 30th, you'll need to be extra cautious on the road. Mechanical problems are possible. This planetary duo will also affect communication, delaying mail and triggering misunderstandings. You're also likely to be more than usually short-tempered and impatient, so remind yourself to think before you speak, write, and act. Take a time-out if you feel frustration rising.

Planetary Lightspots
January's most exciting news is Jupiter, which arrives in Pisces, your solar Tenth House, on the 17th. This beneficial influence can enhance your career and popularity. Plan now to make the most of this expansive planet of good fortune, and target people who can help you soar to new heights.

Relationships
Relationships get back on track after Mercury resumes direct motion on the 15th. Take the initiative to resolve any disagreements that have occurred in recent weeks, and touch base with family members. With Venus entering Aquarius, your solar Ninth House, on the 18th, you'll be in touch with faraway friends and relatives. Why not arrange a reunion?

Money and Success
Saturn in Libra, your solar Fifth House, clashes with Pluto in Capricorn, your solar Eighth House, at month's end. This lineup could impact investments, and this is not the time to risk funds in any speculative venture. There also could be extra expenses related to your children, and company benefits could be reduced or become more expensive. Take precautions to protect valuables and financial information when you're out in public.

Rewarding Days
1, 2, 7, 10, 11, 17, 21, 22, 25, 26, 29

Challenging Days
3, 5, 6, 8, 9, 13, 16, 20, 23, 27, 31

Gemini/February

Planetary Hotspots
Life is upbeat and on track throughout February, with only minor stresses and strains. The exception is Mars in Leo, which is still retrograde in your solar Third House. That can interfere with communication, and the potential continues for mechanical problems as well as mishaps on the road. Avoid difficult people, especially relatives; sparks could fly with an in-law.

Planetary Lightspots
Follow the lead of the February 13 New Moon in Aquarius, your solar Ninth House, and take a short-term class to benefit your career, or attend a conference designed to boost specific skills. The knowledge will begin to pay off as soon as next month. If time permits, take a week's vacation at a relaxing destination where you can recharge for the busy career months ahead.

Relationships
Family relationships are all you could wish for this month, especially around the time of the February 28 Full Moon in Virgo, your solar Fourth House. Connect with relatives near and far, set aside extra time for family activities, and consider hosting a casual get-together at your place the weekend of the 20th. With a little persuasion you probably can get family members to pitch in around the house.

Money and Success
Your star rises as Venus enters Pisces, your solar Tenth House, on the 11th. Even better, the energy continues into March, giving you every opportunity for career success. Organize an after-work outing where you can get better acquainted with coworkers, and watch for the chance to engage decision-makers in casual conversation.

Rewarding Days
4, 7, 8, 11, 13, 16, 18, 26

Challenging Days
2, 3, 9, 17, 19, 20, 23, 27

Gemini/March

Planetary Hotspots
Friendship goes through a rough period when several planets in Aries, your solar Eleventh House, clash with Saturn and Pluto. Money may be the issue, or a coworker could try to undermine your progress. Group activities are also stressful and it will be tough to bring people to consensus. Be wary if you're asked to take on a leadership role.

Planetary Lightspots
Finally! Mars resumes direct motion in Leo, your solar Third House, on the 10th. Communication, your specialty, will slowly begin to get back on track, and people will be more accommodating. Take some time midmonth to reflect on your 2010 goals, the strides you've made, what you can do better, and how best to maximize the year's accomplishments.

Relationships
A relationship will reach a turning point near the March 29 Full Moon in Libra. You and a friend could resolve your differences or part ways, and the same could be true of a dating relationship. But the lunar energy is promising for singles looking for love. Someone new could walk into your life at the least expected moment. Your social life also gets a boost from the Full Moon, so plan a few outings where you can meet people.

Money and Success
You could land a starring role the week of the March 15 New Moon in Pisces, your solar Tenth House—or at least make sizeable career gains. A step up is possible, so go for a promotion or post your resume at an online job site. But the most likely scenario is a sudden opportunity that could be as exciting as it is lucrative.

Rewarding Days
4, 7, 9, 12, 13, 14, 15, 17, 21, 26, 30

Challenging Days
1, 2, 8, 11, 16, 23, 25, 29, 31

Gemini/April

Planetary Hotspots
Saturn retreats into Virgo, your solar Fourth House, on the 7th, and aligns with Uranus in Pisces, your solar Tenth House, on the 26th. Changes involving your career and home are possible, including relocation and company downsizing. It will be difficult to sell property now, at least for a reasonable price, and you should postpone a home purchase until at least this fall. You may also have to deal with challenges regarding a family member, or need a major home repair.

Planetary Lightspots
Smile! You'll be at your most flirtatious and charming after Venus enters your sign on the 25th. This only happens once a year so make the most of it and let your powers of attraction bring you everything you wish for. And go ahead and satisfy your sweet tooth (which is overactive now) once in a while with a favorite treat.

Relationships
The April 14 New Moon in Aries, your solar Eleventh House, energizes your social life and has you in touch with friends and acquaintances. Organize a few group get-togethers and set aside an evening or weekend day for leisurely hours with your best friend. But be sure to confirm plans because Mercury turns retrograde on the 17th. For couples, Venus in Taurus, your solar Twelfth House, accents sizzling romance through the 24th.

Money and Success
April 28 brings the Full Moon in Scorpio, your solar Sixth House. Expect the pace to pick up at work and be sure to double-check projects before you call them complete. This is also your wellness sector, so you'll want to take precautions to avoid a cold or flu. If you have a pet, remember to schedule an annual checkup.

Rewarding Days
1, 2, 3, 9, 10, 11, 13, 14, 17, 27, 30

Challenging Days
4, 5, 7, 12, 19, 23, 25, 28

Gemini/May

Planetary Hotspots

Saturn in Virgo and Jupiter in Pisces align across your solar Fourth/Tenth House axis, which continues to make career and family a balancing act. Be cautious. The more your career responsibilities expand, the more distance you'll create with loved ones. Even they will tolerate this only so long. Think again, however, if a home-based business or telecommuting seems like a good alternative. Long-term, you may regret the decision.

Planetary Lightspots

Uranus enters Aries, your solar Eleventh House, on the 27th. This brief visit, which lasts until mid-August, can lead you to some fascinating new acquaintances, one of whom may become a lifelong friend. But you also can expect at least one friendship to significantly change, probably because your interests are changing.

Relationships

You're drawn to people as they are to you under the May 27 Full Moon in Sagittarius, your solar Seventh House. Emphasize cooperation and compromise, which is what this Full Moon is all about, and learn about yourself as you learn from others. Even so, a relationship will reach a turning point for some Geminis. If you're among them, time apart could help you sort out your feelings.

Money and Success

Venus in Cancer, your solar Second House, from the 19th on has your mind on money. Take charge and take action in financial matters, including budgeting and saving, and curb spending the first half of the month. Unexpected expenses could arise the third week of May or in June, when there's a chance company benefits could be reduced.

Rewarding Days

1, 5, 6, 10, 11, 14, 15, 19, 24, 28

Challenging Days

2, 4, 8, 9, 16, 18, 22, 23, 29

Gemini/June

Planetary Hotspots

Financial stress and strain surrounds the June 26 Full Moon (lunar eclipse) in Capricorn, your solar Eighth House. Pending financial issues will reach a peak that week, and it's possible that your or your partner's income could decline. With Saturn still in Virgo, your domestic sector, be sure property is well covered by insurance. Also check your (and the family's) credit reports and take action if you find any errors.

Planetary Lightspots

The June 12 New Moon in Gemini launches your new solar year with opportunities for personal growth and new directions in the next twelve months. Brainstorm possibilities while Mercury is in your sign from the 10th to 24th. Dream big but also be realistic while you set goals and weigh the pros and cons of various options. Give yourself a fresh start.

Relationships

Jupiter enters Aries, your solar Eleventh House, on the 6th and joins forces with Uranus in the same sign on the 8th. This dynamic combination can trigger sudden luck, probably through a friend, acquaintance, or group; and it's terrific for networking. Do look before you leap, however, and steer clear of anything that requires a significant financial investment.

Money and Success

While tension rises concerning money matters, there's also a potential bright spot. You could luck into a small windfall, and it might be worth investing a dollar in a group lottery purchase. Even so, make spending off limits for all but necessities and build up savings instead. You may need the cushion before year's end.

Rewarding Days

1, 4, 7, 11, 15, 16, 20, 28, 29, 30

Challenging Days

2, 3, 5, 6, 10, 17, 19, 21, 24, 25, 26

Gemini/July

Planetary Hotspots
Saturn leaves your solar Fourth House behind when it returns to Libra on the 21st, and then clashes with Uranus in Aries across your solar Fifth/Eleventh House axis. Keep close tabs on your children if you're a parent, and get to know their friends, especially if you have teens. A friendship or dating relationship could end now or at the least become strained, and you could part ways with a group you've been associated with for many years. Do what's right for you, but only after careful thought.

Planetary Lightspots
Home is your haven this month, with Venus in Virgo, your solar Fourth House, from the 10th on. This is a great time to clean out closets, drawers, basement, attic, and garage. Not only will you get your place in shape but you'll get great satisfaction and instant gratification from tossing out junk.

Relationships
Relationships with relatives will be positive and upbeat this month, and a social event with neighbors could have you in the right place at the right time to make a lucky contact. The same could be true of a community activity, so consider getting involved in a local political campaign or project to improve your city or neighborhood. Meeting people is the ultimate goal.

Money and Success
Finances move into more positive territory this month, thanks to the July 11 New Moon (solar eclipse) in Cancer, your solar Second House. You could find some great bargains on household items during the first ten days of the month, which are also terrific for inexpensive do-it-yourself home improvements. Take your time, though, and be careful on ladders.

Rewarding Days
2, 4, 8, 12, 13, 17, 26, 27, 31

Challenging Days
3, 9, 10, 14, 15, 16, 20, 22, 23, 30

Gemini/August

Planetary Hotspots
An issue that has been building all summer reaches a peak this month when Jupiter, Saturn, Uranus, and Pluto are activated. This four-planet lineup involves your solar Fifth, Eighth, and Eleventh Houses, creating tension and conflict with a friend, group, romantic relationship, and possibly your children. Money may be involved, and you also could discover that someone you thought you knew well is a decidedly different person.

Planetary Lightspots
You need a break. Take one. Aim for a quick trip, a long weekend away near the August 9 New Moon in Leo, your solar Third House. If that's impossible, substitute with a few days off at home to catch up on reading, do puzzles, or play games. Either option will refresh you, body and soul, and give you a much-needed mental respite.

Relationships
Family relationships are generally positive, but be aware that Mercury turns retrograde in Virgo, your solar Fourth House, on the 20th. That can trigger mix-ups and misunderstandings. Make no assumptions, think before you speak, and choose your words with care. Try to discuss important family matters early in the month, but postpone final decisions until later in September if you can.

Money and Success
Your career sector shines brightly under the August 24 Full Moon in Pisces, your solar Tenth House. Use the following two weeks to do all you can to impress decision-makers and to reinforce your position. Uranus returns to Pisces on the 13th, so what you do now could pay off very well next month when Jupiter also returns to Pisces. Set your sights high and listen to the grapevine.

Rewarding Days
1, 5, 9, 13, 14, 17, 22, 23, 28

Challenging Days
4, 6, 7, 8, 10, 12, 18, 19, 20, 25, 27

Gemini/September

Planetary Hotspots
Although the events of the past few months begin to wane, they go one more round in early September and again the week of the September 23 Full Moon in Aries. You may not be able to completely resolve the issues, but you will make progress toward putting them behind you. However, continue to be wary of mixing friendship and money, and also be cautious about investments. Risky moves could bring sizable losses.

Planetary Lightspots
Family life hits the spot under the New Moon in Virgo, your solar Fourth House, on the 8th. Another plus is Mercury, which resumes direct motion in the same sign on the 12th. Take the initiative to clear up any misunderstandings that occurred in the previous weeks, and spend quantity and quality time with family. You'll be in sync and on the same wavelength.

Relationships
Your social life benefits from Venus in Libra, your solar Fifth House, through the 7th, but don't be surprised if you prefer quiet gatherings at home. Once Venus moves on to Scorpio, your solar Sixth House, organize an after-work get-together midmonth with coworkers. Be sure, though, to safeguard valuables when you're out in public.

Money and Success
Jupiter and Uranus join forces once again in Pisces, your solar Tenth House, on the 18th. This lucky influence could again trigger a promotion or job opportunity. If it feels right, negotiate for more money or benefits; you might be successful. Overall, that week promises to be an exciting career week with sudden developments that could thrust you into the limelight.

Rewarding Days
2, 5, 6, 10, 11, 14, 19, 24, 28

Challenging Days
1, 3, 7, 9, 13, 15, 16, 21, 23, 25, 29, 30

Gemini/October

Planetary Hotspots
Venus turns retrograde in Scorpio on the 6th. Placed in your solar Sixth House, you can expect workplace relationships to cool somewhat, which will be true of other ties as well. This is not the time for a commitment decision or wedding. Postpone both until after Venus turns direct in mid-November. The same applies if you want to ask for a raise or need a loan, both of which are unlikely now.

Planetary Lightspots
Tune in when the New Moon in Libra, your solar Fifth House, urges you to explore your creativity. Spend hours with a hobby, using talent and skill instead of big money to create something beautiful and useful. Redo a room, finish a project you began last year, or get started on do-it-yourself holiday gifts.

Relationships
It's early, but your holiday social schedule begins to take off the week of the October 22 Full Moon in Aries, your solar Eleventh House. Line up dates with friends, accept invitations, and consider hosting a Halloween party to get things rolling. Before then, you'll also have meet-and-greet opportunities, thanks to Mercury in Libra from the 3rd to the 19th. Be cautious, though, if a former love interest wants to rekindle the relationship.

Money and Success
Your job is hectic this month, with Mars in Scorpio through the 27th. Keeping up the pace will challenge you at times, but you'll also have the energy and support to meet deadlines if you stay organized. It's well worth the effort because you're sure to be noticed and earn plenty of praise for going all-out.

Rewarding Days
5, 7, 8, 11, 15, 16, 17, 25, 26

Challenging Days
2, 6, 9, 13, 14, 20, 22, 27, 28, 30

Gemini/November

Planetary Hotspots

Life returns to normal this month, at least for the most part. Workplace relationships will be challenging at times because you'll receive mixed messages and be aware that the competition is close behind. But you have luck on your side as long as you don't take on too much. Promise only what you can deliver and then do it in style, showcasing the very best of your skills and talents.

Planetary Lightspots

Time out will be very appealing this month, especially around the November 21 Full Moon in Taurus, your solar Twelfth House. Not that you'll get many hours to yourself. But take as many as you can to rest, relax, and unwind. Meditation can free your subconscious and let your inner voice emerge. Take note of dreams, which can be particularly insightful.

Relationships

Relationships become more easygoing after Venus resumes direct motion on the 18th. People will be more responsive and open-minded, and more willing to grant favors and support your endeavors. If you've been wavering about a dating relationship, your feelings will become clear and you'll either move forward or part ways midmonth. Mercury in Sagittarius, your solar Seventh House, promotes open communication.

Money and Success

November has all the potential to make you a shining star at work, thanks to the New Moon in Scorpio, your solar Sixth House, on the 6th. It could trigger more responsibility and the possibility of a step up. If you're in the market for a new position, send out resumes the week of the New Moon. You could be offered a terrific job before month's end.

Rewarding Days

4, 7, 8, 12, 13, 16, 22, 26

Challenging Days

1, 3, 9, 10, 17, 24, 27, 30

Gemini/December

Planetary Hotspots

Mercury turns retrograde in Capricorn, your solar Eighth House, on the 10th. That can play havoc with payments, so confirm receipt. If you need a loan, or want to change or apply for credit terms, it might be wise to hold off until January; the application could get bogged down in red tape now. Mercury retreats into Sagittarius, your solar Seventh House, on the 18th, where it can trigger misunderstandings with those closest to you. Censor your words before you speak.

Planetary Lightspots

You're in the spotlight of the December 21 Full Moon (lunar eclipse) in your sign. That's a terrific year-end influence because you're prime to set New Year's resolutions and keep them. Don't let anyone deter you from your personal goals, no matter how lofty they are. Once set, almost nothing will stop you from accomplishing them.

Relationships

People will have an important and prominent role in your life this month with several planets and the December 5 New Moon in Sagittarius. Some will surprise you, others will motivate you, and all will inspire you to have faith in yourself. If you're ready to take a relationship to the next level, however, hold off until Mercury turns direct on the 30th, when you'll know if this is the best choice for you.

Money and Success

Despite the potential effects of retrograde Mercury in your solar Eighth House, you'll reap the benefits of Mars in Capricorn from the 7th on. This high-energy planet motivates you to multiply earnings and could trigger a year-end bonus, but you'll want to be very cautious with investments. Postpone major financial decisions until late January if you can.

Rewarding Days

1, 4, 5, 10, 13, 14, 15, 19, 23, 29

Challenging Days

7, 9, 20, 21, 22, 24, 26, 27, 28

Gemini Action Table

These dates reflect the best—but not the only—times for success and ease in these activities, according to your Sun sign.

	JAN	FEB	MAR	APR	MAY	JUN	JUL	AUG	SEP	OCT	NOV	DEC
Move			25, 26		21, 22		10, 11, 15, 16	1–5, 10, 11	16, 17			
Start a class		17, 18	21–31	12, 13		15, 16	12, 13	12, 13			26, 27	
Join a club			21–31	12, 13								
Ask for a raise		24, 25								1		
Look for work			4, 5		25, 26				11, 12			30, 31
Get pro advice		8		3				18				5, 6
Get a loan	13–17			5, 6			22					7, 8
See a doctor			4, 5	15, 16	12, 13			3	24		5	
Start a diet	8, 9		9, 10	28, 29								2, 3
End relationship	11, 12	7, 8	6, 7	3, 4, 30	1, 27, 28							
Buy clothes						19, 20		13, 14				1
Get a makeover	25		21, 22		23–25	11, 12					23	
New romance						19, 20	16, 17	13, 14		6		
Vacation	16–31	1–17										

CANCER

The Crab
June 21 to July 22
♋

Element: Water
Quality: Cardinal
Polarity: Yin/feminine
Planetary Ruler: The Moon
Meditation: I have faith in the promptings of my heart
Gemstone: Pearl
Power Stones: Moonstone, chrysocolla
Key Phrase: I feel
Glyph: Crab's claws

Anatomy: Stomach, breasts
Color: Silver, pearl white
Animal: Crustaceans, cows, chickens
Myths/Legends: Hercules and the Crab, Asherah, Hecate
House: Fourth
Opposite Sign: Capricorn
Flower: Larkspur
Key Word: Receptivity

Your Strengths and Challenges

People perceive you as a nurturer, the one who cares about others and how they feel; and many people benefit from your kindnesses. But some underestimate your inner strength and your ability to withstand adversity. When necessary you take a stand and rarely retreat. When all is safe and secure in your personal world, you're at your best in the outer world. This quest for security motivates you to establish strong family ties and a warm, welcoming home, both of which are central to your contentment and self-confidence.

Ruled by the changeable Moon, you can be upbeat one minute and feel low the next. Only those closest to you, however, realize the depth of your emotions because you have the ability to adapt while using your protective shell to mask your true feelings. You're also highly receptive, intuitively sensing what others miss. But it's wise to be cautious and aware of your environment. You can absorb both negative and positive vibrations from people and your surroundings. Sentimental and loyal, you effortlessly put others at ease and have a long memory for kind and considerate actions. Thoughtfulness is always repaid, even if years later. But your memory bank also stores the negative experiences and you rarely give anyone a second chance.

Your Relationships

Friends have a special place in your heart, and you're devoted to those in your inner circle. Their numbers are few, and you cherish those in whose company you can be yourself. That's when you reveal your inner soul, which you keep hidden from mere acquaintances. You prefer quiet evenings and dinners with close friends to the party scene, and many Cancers are at least somewhat uncomfortable when meeting new people.

Family relationships are among the most important, and you value highly the security that comes with these strong ties. Even if you have issues with family members, your feelings still run deep. For you, family represents the ultimate in relationships. So it's only natural that most Cancers want a family of their own. Remember, though, that children sometimes need to learn the hard way, so let them test themselves in a safe environment.

Finding your soul mate, however, can take a while. It's not so much that you're ultra-choosy as that you instinctively know it's worth the wait to find the right person. But mere passion is not enough to win your heart. You want and need the security of a relationship. And that can make it tough to move on, even when you know it's the best choice for both of you. Many Cancers commit to lifetime love long after their peers, and their patience is usually rewarded with a lasting match. You could click with a Capricorn, your opposite sign, or find love with another water sign, Scorpio or Pisces. Aries and Libra probably aren't the best signs for you, but you might enjoy the practicality of earthy Taurus and Virgo.

Your Career and Money
Whatever your chosen career field, it needs to be one where you can fulfill your leadership potential and where you can exert your independence. Some Cancers excel as entrepreneurs or work in fields such as real estate where they can be their own boss. Teaching appeals to others, but few are happy in a job that requires sitting behind a desk all day. Learning opportunities in the workplace appeal to you, and you do best in a happy, optimistic environment where people and their ideas are valued. Ethical behavior is as important as a paycheck in line with your skills and talents.

Financial security is just as vital to you as emotional security. Home ownership, long-term investments, and savings give you peace of mind, and you're a pro at finding the best bargains, especially on the luxury items you desire. Thrifty no matter how much money you have, some Cancers take this desirable trait too far and become miserly even when their assets are in the millions. If you're among them, remind yourself that money multiplies and returns to you when you're generous with others and yourself.

Your Lighter Side
Few people know the extent of your curious, insightful mind. You pick up what others miss and store all that information for future use. Cancer rarely forget. Of course, part of this talent is your intuition, which helps you link seemingly unrelated facts and events to create a complete picture. With this knack for detail, observation, and analysis you can easily stay a step ahead of the crowd.

Affirmation for the Year
Knowledge is my path to success.

The Year Ahead for Cancer

Lucky Jupiter wraps up its year-long tour of Aquarius when it enters Pisces, your solar Ninth House, January 17. This opens up infinite possibilities for travel, learning, and generally expanding your horizons. Business travel will keep some Cancers on the road, and others will benefit from conferences, seminars, and employee training sessions. You might even be asked to teach an in-house educational program at work. Mid-September could bring a fortuitous contact while traveling, or in any setting where you can showcase your talents. Take advantage of each and every opportunity because what you do this year can pay off handsomely next year in your career.

You'll get a glimpse of what might be in 2011, when Jupiter briefly visits Aries, your solar Tenth House of career and status, June 5–September 7. Make a point to get acquainted with the right people—those who can advance your status—and do all you can to increase your career-related skills and knowledge, since Jupiter spends most of the year in Pisces. If you're unsure how best to proceed, talk with a mentor or someone in your career field. You'll want to be 100 percent prepared with all necessary tools so you're set for major career strides in 2011.

Saturn travels in Libra, your solar Fourth House, most of the year, returning to your solar Third House, April 7–July 20 to conclude its Virgo transit. This gives you an opportunity to complete any Third House matters that might have begun as early as 2007. Take the initiative to resolve any lingering issues with siblings or neighbors. The same applies to any difficulties with a partner or another family member.

Saturn will also make its final exact alignment with Uranus in Pisces across your solar Third/Ninth House axis. Ongoing legal matters may be settled then, although possibly not entirely in your favor. In any case, this is not the most opportune time to initiate legal action nor to put yourself in a precarious position. Drive with care, and don't drink and drive.

Your domestic and family responsibilities will increase with Saturn in Libra. In the next few years, you may need to care for an elderly relative, or find your home life disrupted when an adult child or another relative moves in. This can strain a partnership, so set ground rules, enlist your mate's support, and be open to compromise. On another level, Saturn's Libra transit marks the start of a long-term upward career trend that will culminate about fourteen years from now. Plan accordingly so you're where you want to be when the time arrives.

Uranus in Pisces, your solar Ninth House, echoes this year's knowledge theme, encouraging you to do what's necessary to move up in the world. This planetary placement also encourages you to welcome the wider world into your life. Open your mind to new ideas even if they seem to go against your beliefs. Listen. Read. Research. And then form your own opinions. You might surprise yourself.

Uranus also briefly changes signs this year, visiting Aries, your solar Tenth House, May 27–August 12. This could trigger a career change—anything from a sudden significant promotion to downsizing at your company. In any case you'll see unexpected changes at work, whether they involve you or those around you. That's all the more reason to consider education options.

Be very cautious if you're considering a property purchase. With Saturn in Libra, your solar Fourth House, forming an exact alignment with Uranus in Aries, your timing could be off. And that could leave you with a hefty mortgage on depreciating property. Career relocation is also possible, and you could have difficulty selling your home. But don't postpone necessary repairs. If you do, the problem will only compound itself.

Neptune continues its long transit of Aquarius, the sign it entered in 1998. In the ensuing years you may have already experienced the effects of this inspirational planet on money matters. While it can motivate you to save and invest for the future, Neptune's other side—confusion and illusion—can lead to poor financial decisions. Whatever appears to be a sure thing probably isn't. Ask questions rather than to take anyone's word for it in money matters, and be cautious about investments and whom you contact for advice.

Pluto is now solidly in Capricorn, the sign it will transit until 2024. In the coming years you will experience many relationship changes as people enter and depart your life. Some will be positive and uplifting, encouraging you on a personal level. Others, however, will be decidedly negative—power plays and controlling people. Both will teach you more about relationships and your role in them. As a result you may purposely make a healthy choice to distance yourself from some. In that sense, this Pluto transit through your Seventh House of relationships is more about you than about other people. The best part of this transit is awareness and the courage to do what's right for you.

What This Year's Eclipses Mean for You

Three of this year's eclipses spotlight your solar First/Seventh House axis of relationships, so you can expect other people to have a prominent role in your life throughout 2010. January 15 brings a solar eclipse in Capricorn, followed by a lunar eclipse in the same sign, June 26. An eclipse is active for approximately six months.

The January solar eclipse will trigger an exciting new romance for some singles and other Cancers will form a meaningful friendship or find a mentor. Knowledge—learning from others and doing the same for them—is also tied to this eclipse. So don't be surprised if someone suddenly appears in your life at just the right time to guide you in a new direction.

But what begins as a positive relationship may become bumpy or possibly end during the second half of 2010 because the June lunar eclipse activates Pluto, Mercury, Jupiter, and Uranus. You could realize that someone you previously saw as uplifting has a less attractive side, and that the relationship is primarily for someone else's gain, not yours. This planetary combination can also indicate obsessive thinking (you or someone else) and a skewed viewpoint. Try to remain objective. If you hold on too tightly—to another person or a coveted dream—you could block the energy from working in your favor.

And you'll very much want to clear the way for the future on a personal level because the July 11 solar eclipse occurs in your sign. It might finally be time to put yourself and your needs first on your priority list. Thinking and planning are parts of the process because the eclipse activates Mars in Virgo, your solar Third House of communication and learning. A short-term class could benefit your career, or you might want to browse the self-help section for books that can help you better define your place in the world and what you want from life.

December 21 brings the fourth eclipse of the year—a lunar eclipse in Gemini, your solar Twelfth House of self-renewal. This eclipses reinforces the process you began at the July solar eclipse, encouraging you to look within for answers. Listen to your sixth sense, which will be active, especially concerning career matters. Watch and wait for an opportunity that's likely to pop up when you least expect it, possibly while traveling on business. But with retrograde Mercury also involved in the eclipse, you'll want to let matters unfold in their own time and get everything locked in before you make a move. Be especially cautious wherever money is involved, and do the same with legal matters; the outcome is likely to be other than what you expect.

Saturn

You'll find it easy to tap Saturn's steadying influence **if you were born between July 20 and 22**. This serious planet will contact your Sun from Virgo, your solar Third House, between April 7 and July 20, bringing with it added determination to accomplish whatever you focus on. Saturn here sharpens your thinking and enhances your organizational ability. Learning will be a strong theme, and you might want to consider returning to school; or, take a class for the fun of it or to increase your job marketability. You'll also learn from other people, and someone may have a profound impact on your life, providing the spark to make this year a personal or professional turning point. Serious thinking accompanies this transit, and you may have a tendency to worry far more than usual. Rein in your mind and replace those thoughts with practical, realistic, upbeat ones. Remember, negative thinking can spark negative events and positive thoughts can bring abundance. You may also find yourself thinking more about the past. If regrets emerge, resolve them by taking action, including making amends with people—friends, partner, relatives, and even neighbors.

It's also possible you'll need a new vehicle, especially if this is something you've postponed. Significant repairs are also possible, and it would be wise to weigh the cost against the return. Sometimes it's better to cut your losses and begin anew.

If you were born between June 21 and July 8, Saturn will clash with your Sun from Libra, your solar Fourth House. Saturn's influence will be minor, however, **if your birthday is after June 26**—a few weeks at the most between August and year's end. Cancers born in the first period will feel Saturn's effects from January 1 to April 6, and again from July 20 to early August.

With serious Saturn in your domestic sector, you can expect home and family to be more of a priority, along with increased responsibilities. You may have to manage the affairs of an elderly relative, or there may be changes in your living situation, with a family member moving in or out. It's also possible your home may need significant repairs or that you could relocate. Saturn here also emphasizes possibilities. Roll up your sleeves and get busy if your home needs a facelift or fresh paint and new décor, and your efforts will be well rewarded.

Be cautious, though, if you're considering a property purchase or a new roommate, and postpone either until next year if you can. A promising relationship could end before then, or a job or family event could make you wish you were without a mortgage payment. Financial partnerships are in the same category. Don't let someone use your good credit for a major purchase; you could end up with the debt.

This Saturn transit is important for another reason: it marks the beginning of a career cycle. So although the domestic scene will be your main focus, you'll also feel the pull of the outer world. Give some thought to where you want to be in the future and then develop short- and long-term goals that will help you achieve your objective. Open your mind to possibilities rather than focusing on your current career; the time could be right to begin to pursue another avenue.

Uranus

Uranus will contact your Sun from Pisces, your solar Ninth House, **if you were born between July 15 and 22**. This favorable connection encourages you to invite change into your life—on your terms. Do a self-assessment from head to toe, from brains to brawn, from your deepest wishes to your worldly goals. Then take charge and tackle your personal to-do list. Be sure to include education on your list because what you learn now will begin to pay off next year. If you're nearly finished with a degree, keep pushing yourself. Success is in sight with only one more hurdle to jump in April when Saturn and Uranus form their final Virgo/Pisces alignment.

Uranus in your solar Ninth House also encourages you to look to the future, to imagine possibilities. You have that insight now and can mould yourself to achieve almost anything. Mid-September is particularly favorable, so you'll want to be aware and alert to sudden opportunities, possibly while traveling. Apply then if you need a scholarship or have your sights set on an exciting new job.

If you were born June 21 or 22, you're among the first of your sign to have Uranus contact your Sun from Aries, your solar Tenth House of career and status, May 27–August 12. Uranus could trigger a sudden career change—a step up, new job, or unfortunately, downsizing. But with Jupiter merging its energy with Uranus at the same time, the odds are you'll not only land on your feet but ultimately be in a far better position. Keep in mind, though, that Jupiter also has a reputation for undelivered promises, and it could thus be next year before you see the true benefits of this transit.

As lucky as Jupiter can be, it's also important to weigh the potential downside of Saturn in Libra, which will align with Uranus in Aries across your solar Fourth/Tenth House axis. With this lineup you'll want to think carefully about a home purchase. Relocation is more than possible. Also, be sure your home is fully insured, especially if you live in an area prone to severe weather. If you rent, insure your possessions.

Neptune

If you were born between July 16 and 22, Neptune will contact your Sun from Aquarius, your solar Eighth House of joint resources. Although this potentially stressful contact can trigger financial worries, the real reason this transit is stressful is because it will be difficult to get a handle on money matters. Be proactive, and save rather than spend. Be conservative with investments and postpone all but the most necessary major financial decisions. The good news is that Neptune is known to provide exactly what you need when you need it. Nevertheless, it would be unwise to enter into a financial liaison with anyone this year, no matter what you're told or promised.

Pluto

Pluto will contact your Sun from Capricorn, your solar Seventh House, **if you were born between June 24 and 26**. Although this is a difficult transit that occurs only once in a lifetime, it does have its upside. First, the tough part. Pluto transforms whatever it touches. This means you and someone close to you will undergo irreversible change in the coming year: the relationship could deepen or end. (There is no in-between with Pluto.) Or you could suddenly realize you're caught up in an obsessive relationship.

What Pluto challenges you to do is to examine not only the relationships in your life but your perspective on relationships, and then make the necessary changes. This process probably won't be an easy one, but you'll emerge stronger and wiser, which is Pluto's ultimate aim. And nearly everyone who undergoes a Pluto transit is a better, more confident person for the experience.

You can also use Pluto to revamp yourself through diet, exercise, and healthy eating and living. Use this powerful planet to replace bad habits with good ones.

Jupiter-Saturn-Uranus-Pluto
If you were born between June 21 and 23, this intense four-planet alignment will contact your Sun this summer. Change is a given with this lineup, and it's unlikely to be stress-free. More to the point, you will find the summer months to be both stressful and uplifting. You'll look to the future and unlimited opportunities while also clinging to the past. Letting go and embracing change is thus your major challenge with this alignment.

Forecasting possible events associated with these planets is a guessing game at best, with Uranus involved. Nevertheless, you can expect events to involve home and family, close relationships, your career, and, of course, you. Some Cancers may end a relationship, make a major move, or change careers.

But it's important to first talk with yourself and specifically identify your wants, needs, desires, and motives. Because ultimately this is about you, not other people, your home, or your career. They are simply the outward manifestation of changes going on within you. One final word: the changes you make now will be permanent, so choose wisely.

 # Cancer/January

Planetary Hotspots
Domestic and family matters are this month's hotspot as Saturn in Libra, your solar Fourth House, clashes with Pluto in Capricorn, your solar Seventh House. This can trigger difficulties among family members or between you and your mate. Talk things out after establishing ground rules and the importance of compromise and trust. It's also possible your home will need a major repair, or that you'll consider relocation. This is not the month to welcome a new roommate.

Planetary Lightspots
Get ready, get set, go! Your year of adventure begins January 17, when Jupiter enters Pisces, your solar Ninth House. Travel will be on your agenda, along with the desire for knowledge. Do both. Take a fabulous trip where you can learn about other cultures. Also consider short-term classes or a one-year program that can advance your career or prepare for a new one you can launch next year.

Relationships
The January 15 New Moon (solar eclipse) in Capricorn highlights your relationship sector, putting you in touch with old friends and new faces this month. With Mercury turning direct in Capricorn on the 15th, and Venus in the same sign through the 17th, you'll attract many people into your life. These planetary events will also help you see the other person's viewpoint.

Money and Success
With Mars retrograde in Leo, your solar Second House, anticipated income could be slow in coming. A raise or bonus may be put on hold, and there can be mix-ups with bills and payments. Check statements carefully when they arrive and pay bills early. Mars's effect will be more pronounced at month's end under the Full Moon in Leo on the 30th, and carry over into the first two weeks of February.

Rewarding Days
1, 4, 8, 9, 14, 18, 19, 25, 28

Challenging Days
5, 6, 13, 16, 20, 21, 27, 31

Cancer/February

Planetary Hotspots
Life rolls along in February with the exception of Mars, which is still retrograde in Leo, your solar Second House. Continue to watch expenses, especially your household budget, and hold off on any big-ticket domestic purchases if you can. Get a second and even a third quote if a repair is necessary, and also take time to check references.

Planetary Lightspots
Start thinking about a spring vacation destination after Venus enters Pisces, your solar Ninth House, on the 11th, either with your mate or the family. Traveling singles could make a match out of town, but a long-distance relationship probably won't be long-lived. In the meantime, enjoy it for what it is.

Relationships
You're on the same wavelength with just about everyone in your circle as Mercury advances in Capricorn, your solar Seventh House, through the 9th. If you're part of a couple, set aside time to catch up with your partner's life. You might be pleasantly surprised by what you hear regarding his or her thoughts about the future. Someone at a distance brings you luck midmonth.

Money and Success
The February 12 New Moon in Aquarius, your solar Eighth House, is a positive financial emphasis that could trigger a gift from a family member or a minor windfall. But it also might be necessary to apply for credit or a loan to finance a household item. Shop around in search of the best interest rate and have faith that all will come together. The odds are it will, especially if you do the necessary research. Avoid snap decisions in money matters, ask questions, and read the fine print.

Rewarding Days
1, 4, 5, 6, 10, 14, 16, 24, 25

Challenging Days
2, 3, 9, 17, 19, 20, 21, 23

Cancer/March

Planetary Hotspots
March brings action in your Aries solar Tenth House, along with career-related stress and conflict. A power play is likely, and personality conflicts will disrupt progress. Be cautious here. A clash with the boss could have unwanted consequences. Stand your ground, but with finesse, and be sure to document talks and actions. If your company has an employee dating policy, be sure to follow it rather than put your job at risk.

Planetary Lightspots
Knowledge will be your guiding light this month, as several planets and the March 15 New Moon in Pisces spotlight your solar Ninth House. Your mind will be extra sharp and ready to absorb any and all information. Listen to your intuition. Fresh insights of understanding about the world at large and your place in it will grab your attention at unexpected moments.

Relationships
Family ties will strengthen under the March 29 Full Moon in Libra, your solar Fourth House, and you'll delight in spending as many hours as possible with loved ones. Be sure to touch base with out-of-town relatives in early March, and give some thought to inviting them for a brief visit. But think twice about a trip to your hometown; it could trigger regrets.

Money and Success
Money matters begin to regain momentum after Mars resumes direct motion in Leo, your solar Second House, on the 10th. Although it will be April before things are back to normal, you can use this time to organize financial records, review and revise budgets, and create a plan to maximize income during the rest of the year. You might get a great deal on a household item if you shop the third week of March.

Rewarding Days
4, 5, 9, 10, 13, 14, 15, 19, 27, 28

Challenging Days
1, 2, 8, 11, 16, 20, 21, 23, 25, 29

 # Cancer/April

Planetary Hotspots
Saturn retreats into Virgo, your solar Third House, for its final few months in that sign. There, it aligns with Uranus in Pisces, your solar Ninth House, on the 26th. This is a positive influence if you're completing a degree or advanced course of study. If not, you still have time to gain the extra knowledge that can benefit your career. Mechanical problems with a vehicle or appliance could pop up now, and any ongoing legal matters will take a step toward resolution. If you have difficulties with a neighbor or relative, do all you can to smooth things over.

Planetary Lightspots
Catch some spring fever! With the April 28 Full Moon in Scorpio, your solar Fifth House, there's no time like now to put your to-do list on hold and get outside. Play, see friends, take your kids somewhere fun, and, if you're single, launch a whirlwind romance. You'll also be in meet-and-greet mood, so socialize with friends where you can connect with new faces.

Relationships
Mercury, which turns retrograde on the 17th in Taurus, your Eleventh House, can trigger misunderstandings with friends. Try to prevent that from happening, and also confirm dates and times so you don't miss out on a social event. Someone you meet or talk with in early April could turn out to be a terrific networking contact mid-May.

Money and Success
You're in the spotlight, a real attention-getter, thanks to the April 14 New Moon in Aries, your solar Tenth House. Make a point to talk with all the right people and to subtly promote your skills and talents. These efforts could lead to rewards this summer.

Rewarding Days
1, 3, 5, 6, 10, 11, 16, 20, 21, 29

Challenging Days
4, 7, 12, 15, 19, 23, 25

 # Cancer/May

Planetary Hotspots
The alignment of Saturn in Pisces and Jupiter in Virgo across your solar Third/Ninth House axis highlights travel and learning. Although a trip may be necessary to assist a relative, possibly an in-law, the main focus of this planetary duo is inward. Use this time to reflect on the past and learn from it; resolve regrets, make amends, seek advice. The result will be the best of Jupiter's optimism for the future combined with the knowledge of Saturn.

Planetary Lightspots
Need a reason to briefly step out of your hectic life? You have it this month with Venus in Gemini, your solar Twelfth House, through the 18th. Spend a little time every day doing what you most enjoy. Also try meditation, which can enhance your sixth sense and trigger fresh insights into your life, the past, and the future.

Relationships
Fun and friendship are the ideal mix under the May 13 New Moon in Taurus, your solar Eleventh House. And with Mercury resuming direct motion in Taurus on the 11th, your social life will pick up speed. Take note of anyone you meet during the first few days of May; this person could surface again in mid-May as an excellent networking contact.

Money and Success
Uranus enters Aries, your solar Tenth House on the 27th, where it will remain through mid-August. Look, listen, and use your intuition to catch a hint of possible career changes that might develop this summer. The 27th also brings the Full Moon in Sagittarius, your solar Sixth House of daily work, and a highly productive period that can put you a step ahead of the competition.

Rewarding Days
3, 7, 8, 12, 17, 21, 24, 25, 30, 31

Challenging Days
2, 4, 9, 14, 16, 18, 22, 23, 27, 29

Cancer/June

Planetary Hotspots
Your solar First House is this month's hotspot, primarily because you'll feel an increasing desire and need for change. That can be both positive and negative, depending upon the timing and the decisions. Premature action can backfire and cause difficulties with people close to you. But it's also possible that career conditions and relationships could trigger a snap decision. Be cautious. Things will evolve throughout the summer.

Planetary Lightspots
Finances are promising this month, with both Venus and Mars visiting Leo, your solar Second House of personal resources. This puts you in a good bargaining position for a raise or additional benefits midmonth—if your sixth sense agrees. That timing also favors shopping for career clothing; you could find some sensational buys.

Relationships
You'll experience nearly every facet of relationships at month's end when the June 26 New Moon (lunar eclipse) in Capricorn activates your solar Seventh House. Uplifting moments are in the forecast, but so is conflict. Do your best to avoid controlling people and try to view situations with detachment rather than letting frustration get the best of you. Be careful about whom you trust.

Money and Success
Jupiter enters Aries, your solar Tenth House, on the 6th, and merges with Uranus in the same sign two days later. This lucky lineup could bring the first hint of a fantastic career opportunity. But it may come with some risk, so you should get all the facts and weigh them against potential rewards. In addition, things could change rapidly in the next few months, so you might want to hold off on a decision until this fall.

Rewarding Days
4, 5, 8, 9, 13, 14, 17, 18, 22, 27

Challenging Days
3, 6, 19, 21, 24, 25, 26

 # Cancer/July

Planetary Hotspots

Saturn returns to Libra, your solar Fourth House, on the 21st, and aligns with Uranus in Aries on the 26th. This indicates topsy-turvy career conditions that can land you on top; but with Uranus involved, company downsizing and relocation are also possibilities. No matter how great the deal, this isn't the best time to purchase a home or invite a roommate to share your space. If you have a home to sell, you could be successful but realize only a small gain or no gain at all.

Planetary Lightspots

Get set to thrive! That's the message of the July 11 New Moon in your sign, which signals the start of your new solar year. Tap into the fresh lunar energy and set an ambitious path of personal achievement and new directions for the next twelve months. You'll get the benefit of Mars (aligned with the New Moon) and all the initiative and incentive it represents. Brainstorm your way to success.

Relationships

Relationships benefit from open communication as Venus travels in Virgo, your solar Third House, from the 10th on; it's joined by Mercury on the 27th. Listen between the lines because you may gain more from what's left unsaid than from what you hear. A mentor or partner will offer wise advice at month's end.

Money and Success

The July 25 Full Moon in Aquarius, your solar Eighth House, accents money matters—lenders, insurance, investments, and your partner's income. Be particularly cautious with investments and take no one's word on faith in financial matters. If you need a loan, try to hold off for a few months; but if it's a must, be sure to read all the fine print.

Rewarding Days

1, 6, 7, 11, 14, 15, 19, 24, 28, 29

Challenging Days

3, 4, 5, 10, 16, 20, 22, 23, 30

 # Cancer/August

Planetary Hotspots
Recent developments involving your home and family, career, and relationships come to a peak this month as Jupiter, Uranus, Saturn, and Pluto align in your solar Fourth, Seventh, and Tenth Houses. Changes in all these areas of life are possible, even likely, and events will unfold quickly during the first three weeks of August. Be sure your home and property are well insured, especially if you live in a severe weather area, and also check in with relatives. Again, you may experience career-related changes, including a promotion, new position, downsizing, or relocation.

Planetary Lightspots
Take advantage of Venus in Libra, your solar Fourth House, after the 5th, to move through your home, room by room. Clean every closet and storage space, toss out what you don't need, take other things to a consignment shop, hold a yard sale, and generally get your place in shape. Although time will be at a premium this month, redo a room if you can manage it. You'll love the results.

Relationships
Some relationships are tension-filled as several planets clash with Pluto in Capricorn, your solar Seventh House. Others are prone to misunderstandings with Mercury turning retrograde in Virgo, your solar Third House of communication, on the 20th. Clarify your thoughts and words in the next three weeks rather than assuming people understand where you're coming from.

Money and Success
The August 9 New Moon in Leo, your solar Second House, casts its beams on money matters, income and spending. Although extra money could flow your way under this influence, status quo is more likely. Do be careful, however, with personal financial information.

Rewarding Days
2, 3, 7, 8, 11, 15, 16, 21, 24, 29, 30

Challenging Days
1, 5, 6, 13, 14, 17, 18, 22, 23, 27, 28

 # Cancer/September

Planetary Hotspots

You can expect periodic tension in family relationships as several planets transit Libra, your solar Fourth House. A stressful career- or home-related event could be the issue, as could the arrival of a relative or the need to relocate. Try to curb your impatience with loved ones by finding a constructive outlet for frustration. Talk is productive if you listen and open your mind to compromise.

Planetary Lightspots

Both your solar Third and Ninth Houses are in focus this month, and that's all the reason you need to get away for a week or a weekend or to expand your knowledge. The Jupiter-Uranus merger in Pisces is great for business travel, and the September 8 New Moon in Virgo might be just the incentive you need to take a class for the fun of it or to master a specific skill.

Relationships

Your solar Fifth House comes alive on the 8th when Venus enters Scorpio, followed by Mars on the 14th. That's a terrific influence for everything from romance to socializing to time with your children. If you're single, you could meet someone fascinating midmonth. The same timing favors sporting activities, so consider joining a gym. Start slowly, though. Mars can encourage you to overdo it.

Money and Success

Career stress accompanies the September 23 Full Moon in Aries, your solar Tenth House, when the lunar energy activates Jupiter, Saturn, and Pluto. The good news is this is the last major round of the planetary lineup that's been active all summer. Tread carefully and try not to step on toes, while also doing all you can to avoid being drawn into a power play. Again, you could rise to the top or feel the stress of company downsizing.

Rewarding Days

6, 7, 8, 10, 11, 12, 17, 20, 22, 27

Challenging Days

1, 3, 4, 9, 15, 16, 23, 25, 30

 # Cancer/October

Planetary Hotspots
Venus turns retrograde in Scorpio, your solar Fifth House, on the 8th, and stays that way through mid-November. Hold off if you're considering a partnership (business or romantic) because you could have a change of mind or heart. Dating relationships are unlikely to click now, and people in general will be more distant. You'll also want to be cautious with investments and postpone major financial decisions and purchases.

Planetary Lightspots
Despite the influence of retrograde Venus, you'll still have plenty of opportunities to socialize this month, thanks to other favorable planetary alignments in Scorpio. Romance sizzles for couples in early October and communication is at its best at month's end, when a relaxing vacation trip is in the forecast for some.

Relationships
Lingering influences from this summer's planetary alignments will affect family communication at times. Overall, though, these relationships will be upbeat and inspirational, and you'll have time to reflect on their importance in your life. If you've never explored your family history, this is a great month to begin that project. It could ultimately lead to meeting relatives in another part of the world.

Money and Success
The year's second Full Moon in Aries, your solar Tenth House, occurs on the 22nd. This one, however, will be a bit calmer. It's also a Full Moon that encourages you to wrap up any unfinished career-related business, projects, and decisions. That's also a wise idea on a practical level because November will be a very busy month on the job.

Rewarding Days
1, 4, 5, 9, 10, 18, 19, 24, 29

Challenging Days
2, 6, 7, 13, 14, 17, 20, 23, 27

 Cancer/November

Planetary Hotspots

Life more or less returns to your normal hectic pace this month, with only minor challenges. Among them is the possibility of a personality conflict at work, and increased potential for complications if you travel during the Thanksgiving holiday. It's also possible you could lose a treasured or valuable possession, so be sure to take precautions when you're shopping, socializing, or traveling.

Planetary Lightspots

Don't be surprised if you get a sudden urge to redecorate as Venus resumes direct motion in Libra, your solar Fourth House, on the 18th. Curb your enthusiasm just a little so you don't dive in without a plan, or plunge ahead with an overly ambitious one. Prioritize and set a goal to complete one project by month's end.

Relationships

Your holiday social life moves into high gear, thanks to the November 6 New Moon in Scorpio and the Full Moon in Taurus on the 21st. Make dates, host a party, accept invitations, mix and mingle. You're sure to meet some fascinating people this month, either in town or while traveling. If you're looking for a new relationship, ask friends if they know anyone who might interest you.

Money and Success

With Mars in Sagittarius, your solar Sixth House, you can expect a hectic month at work, and even more so once Mercury enters the same sign on the 8th. Organization is a must in order to complete your to-do list because you'll have many interruptions and distractions. But you'll also end the month with great satisfaction even if you do feel a little harried.

Rewarding Days
1, 2, 4, 5, 12, 14, 16, 19, 20, 28

Challenging Days
3, 6, 9, 10, 15, 17, 23, 24, 27, 30

Cancer/December

Planetary Hotspots

Mercury turns retrograde in Capricorn, your solar Seventh House, on the 10th. That can trigger multiple misunderstandings that lead to conflict when Mars, which enters the same sign on the 7th, clashes with several planets. Communication and compromise will help ease tension in your personal life, and on the job from the 18th on, the date Mercury retreats into Sagittarius, your solar Sixth House.

Planetary Lightspots

Chances are, you'll need a few days off this month to rest, relax, and get away from the daily grind. Aim for the week of the December 21 Full Moon (lunar eclipse) in Gemini, your solar Twelfth House of self-renewal. Forget the outer world, kick back, and enjoy time with yourself, family, and friends.

Relationships

Relationships also have their upside this month, thanks to Venus in Scorpio, your solar Fifth House. That's a great influence for casual get-togethers with friends and memorable moments with your children. If you're wavering about a serious dating relationship, however, this month's planetary alignments could prompt you to move on. Take the time to get better acquainted with coworkers; one of them could provide just the lucky connection you're looking for.

Money and Success

Be sure to check and re-check projects and routine work after retrograde Mercury enters Sagittarius. Also be sure to document important instructions. This way, a major mix-up or error is less likely to impact potential job gains. Keep in mind that final decisions may be reversed when new information comes to light. Protect yourself by re-reading e-mail before you hit send.

Rewarding Days

2, 3, 8, 12, 13, 16, 17, 23, 25, 30, 31

Challenging Days

1, 6, 7, 9, 14, 15, 20, 21, 22, 27, 28

Cancer Action Table

These dates reflect the best—but not the only—times for success and ease in these activities, according to your Sun sign.

	JAN	FEB	MAR	APR	MAY	JUN	JUL	AUG	SEP	OCT	NOV	DEC
Move		2–4				19, 20		13, 14				
Start a class	4, 5	1			21, 22		28–31	2–5	16–22		1, 2	
Join a club	23, 24			5–16	12–19			2, 3				
Ask for a raise			25, 26			15, 16					26	
Look for work	11, 12			3						11, 12	22–30	4–6
Get pro advice	18, 19	9	9, 10	6	31				16, 17			7, 8
Get a loan	21, 22, 25, 26	8, 12										
See a doctor	25, 26	23	22		14	11					23	
Start a diet	11, 12	7, 8	6, 7	3, 4	27, 28							
End relationship	13, 14	9, 10	9, 10	5, 6	30, 31	26, 27						
Buy clothes	8, 9		4, 5						11, 12			3, 4, 7, 8, 31
Get a makeover		24, 25				13	11			1		
New romance					25, 26		20		11, 12, 16, 17, 22, 30	1, 2		
Vacation		19–28	1–6									

The Lion
July 22 to August 22

♌

Element: Fire
Quality: Fixed
Polarity: Yang/masculine
Planetary Ruler: The Sun
Meditation: I trust in the strength of my soul
Gemstone: Ruby
Power Stones: Topaz, sardonyx
Key Phrase: I will
Glyph: Lion's tail

Anatomy: Heart, upper back
Color: Gold, scarlet
Animal: Lions, large cats
Myths/Legends: Apollo, Isis, Helios
House: Fifth
Opposite Sign: Aquarius
Flower: Marigold, sunflower
Key Word: Magnetic

Your Strengths and Challenges

Generous, loyal, confident, and outgoing, your sunny personality brightens the world with a cheerful, can-do optimism. Honesty is another of your sterling qualities, and one that's an ideal match for your natural leadership ability that plays well on center stage. The limelight is your preferred location, where you shine with all the royal presence your sign is noted for.

But with so much attention focused your way, it's easy to forget that others also need applause and the accompanying ego boost. Remember that. Share the credit, be the first to praise others for their accomplishments, and set an example as a motivational team player who values even the smallest contribution.

Sun-ruled Leo is a fixed-fire sign, which means you have the best of both worlds: you're a starter and a finisher, the one who can get things moving and then complete the task. The downside, however, is that you're also stubborn at times, unwilling to compromise when that might be to your advantage. But that quality can just as easily be expressed as determination, which is a positive that will get you much further in life. You take much pride in your reputation, and your self-esteem is tied directly to what others think. So Leos rarely act in ways they'll later regret. You're also among the most popular and live life to its fullest, approaching each day with enthusiasm and a winning attitude. Focus on that when you feel slighted, when your sensitive soul is bruised.

Your Relationships

Surrounded by a wide, ever-changing circle of acquaintances, curiosity motivates you to be the first to meet someone new. You delight in impromptu gatherings and lively conversation, and the grand entrance is just your style. But you also enjoy quieter moments when you can kick back and enjoy long talks with those few you call close friends.

Few can match your flair for the dramatic when it comes to romance. You know how to impress a date (and later, your mate), and excel at playing the field. Each new relationship is an adventure, a new personality to explore, another heart to win. You're perfectly content with that scenario until you meet someone with partnership potential. But he or she might be tough to catch because you're attracted to independent types. That's the influence of Aquarius, your opposite sign and one that needs a high degree of freedom and autonomy. Love could sizzle with one of the other fire signs, Aries and Sagittarius, but sharing the spotlight with another Leo might be tough. Taurus and Scorpio may be too possessive for you.

Children bring out your playful spirit, and most Leos want a family. As a parent, you can be overindulgent, but you also stress the importance of values and education and give your children every possible opportunity to explore their skills and talents. You're their cheerleader, their biggest fan, and you encourage them to embrace life with the same optimism that guides you.

Your Career and Money

One life-long career is your desire. But that may be unrealistic, so keep your options open and be aware of opportunities in related careers that might provide the security you desire. And once you set your sights on a higher position, you persist until success is realized. In any career you can be a leader and a rock, the one who promotes productivity and keeps things humming.

You're task-oriented and ambitious, which can be both a plus and a minus. Although that motivation can help you advance, you also have a tendency to push yourself and others too hard. Keep things in perspective, and remember that not everyone shares your desire to climb the ladder. A big-picture financial view is also important, because you can get wrapped up in the details and neglect the long-term. You excel at budgeting, and you can build the assets necessary for a comfortable lifestyle and retirement. Remind yourself of this when you want to donate more than you can realistically afford in support of a charitable cause.

Your Lighter Side

People see you as confident and outgoing, which you definitely are. What most don't realize, however, is that you're also intuitive and thus often sense undercurrents that others miss. You also have a sixth sense about people. Develop this ability through meditation or simply sit back, quiet your mind, and observe your environment and the people in it.

Your Affirmation for the Year

My action plan is financial security.

The Year Ahead for Leo

Wishing for riches? They could be yours this year, thanks to Jupiter in Pisces, your solar Eighth House. Keep in mind, though, that Jupiter here can increase debt as much as it can boost income. Make a pact with yourself to live within your means, to save and invest, and to pay any accumulated debt before you spend. This way you'll wrap up 2010 with a healthier balance sheet than you had in January. It also can be tempting to indulge loved ones or to romance your latest love interest with only the best of the best. Restrain yourself. After all, Jupiter's largesse won't last forever and a habit formed is tough to break.

Jupiter briefly visits Aries, June 5–September 7, giving you a preview of 2011. Placed in your solar Ninth House, travel may be on your summer schedule with more to come next year. Aries is also your knowledge sector, so don't be surprised if you have sudden urge to learn more about a subject that's always interested you. Enroll in summer school or take a class for the fun of it just to whet your appetite; more in-depth study is on tap for next year.

Spirituality is also associated with the Ninth House and Jupiter, so you'll also feel this part of yourself come alive. Expect more questions than answers during this initial Jupiter transit in Aries that also encourages you to open your mind to new ideas and possibilities.

Saturn begins and ends the year in Libra, returning to Virgo April 7–July 20 to wrap up unfinished business. Financial matters have been at the forefront since 2007, when Saturn entered Virgo, your solar Second House of personal resources. This has undoubtedly been a learning experience, which is Saturn's hallmark, as well as one that gave you (and continues to give you) the opportunity to develop new spending and saving habits. If not, Saturn offers you a second chance this year to accomplish this task. If you've already realigned your financial thinking, you'll want to revisit all that you've accomplished and learned and get the best from Jupiter in your solar Eighth House.

Saturn in the Second House can restrict resources, emphasizing the need for a tighter budget. Income can also decline with this transit. Above all, however, Saturn here challenges you to redefine what you value. What's more important? A home theater system or money in the bank? Adding more to your retirement account or a luxury vacation?

Saturn in Libra spotlights your solar Third House of communication and quick trips. This is also your learning sector and the area associated with siblings, neighbors and the neighborhood, and extended family. So you can expect all of these people and activities to have a more prominent place in your life the next few years.

Known as the planet of karma and responsibility, Saturn is the great teacher, bringing lessons to master. Although Saturn encourages book-learning as a path to worldly success, especially in the Third House, this planet focuses on experiential learning. This situation provides opportunities to expand your communication and people skills. You'll also gain through listening because people will share information. So keep an open mind and absorb all you can.

Saturn also represents the past. Memories will thus occupy your thoughts along with some regrets for what might have been. This is all part of Saturn's role in helping you resolve lingering issues that can hold you back.

Another excellent use of this Saturn transit is education. You'll have the mental focus and stamina to stick with a course of study. The same is true if you've ever wanted to write a book or research your family history. You might also want to get involved in your community or neighborhood association.

Uranus also visits two signs this year, spending most of its time in Pisces, your solar Eighth House of joint resources. Placed here, Uranus can trigger a windfall as easily as unexpected expenses. Family income can rise and fall, and this is not the time to make risky investments.

But like Saturn in the Second House, Uranus in the Eighth House encourages you to focus on financial values in terms of debt, retirement funds, and getting the most for your money. It also emphasizes the importance of savings to weather the ups and down of Uranus. This continues to be vital because again this year Saturn in Virgo clashes with Uranus in Pisces. Plan ahead and save and live within a conservative budget so you're prepared if this planetary duo plays havoc with your finances.

Uranus briefly visits Aries, your solar Ninth House, from May 27 to August 12. This is your travel and knowledge sector, and both will be a part of your life for about the next seven years. Uranus here reinforces Saturn's Third House message that it may be time to return to school to begin or complete a degree, or to get specialized training to advance in your career field. You may see the wisdom of this option this summer when Saturn forms an exact alignment with Uranus. Explore your options so you're set to go in earnest next year when Uranus returns to Aries.

This summer may also bring unexpected travel, possibly associated with a family member. Or you could attend a reunion and reconnect with someone from the past who can open doors. But think carefully if you're considering legal action; the end result may not be what you want or expect. And take no chances on the road. As always, go with a designated driver or catch a taxi.

Neptune continues its long transit of Aquarius, your solar Seventh House, which it entered in 1995. Having the ultimate romance planet in your partnership sign can be as good as it gets. But Neptune is also the planet of illusion and confusion, which can make it tough to know love from infatuation. This transit is not limited to romance, however. It applies to business partners and professional consultations, as well as some friends and coworkers. Be careful whom you consult (attorney, physician, CPA, etc.) and be sure to check references and qualifications. Also be skeptical if you're told only what you want to hear or if anyone promises or guarantees results.

Neptune can also link you with positive, uplifting people who can be just the inspiration you need. Do the same for others.

Pluto will trigger many changes in your work life during its long transit of Capricorn. This slow-moving planet, which will be in your solar Sixth House until 2024, is noted for change on a global scale as well as a personal one. In the coming years you're likely to see your workplace undergo sweeping changes, possibly because of technological or financial conditions. Some jobs, although not necessarily yours, may become obsolete or require complete retraining. This is Pluto's transformative effect. And even though it can be a difficult process, the end result is usually positive.

The Sixth House is also your health and wellness sector. So do yourself a favor and get routine medical, dental, and eye checkups. You also can use Pluto's influence to transform your lifestyle into a healthier one: nutritious diet, exercise, sleep, and relaxation. Tap into Pluto's willpower.

What This Year's Eclipses Mean for You

Your solar Sixth House is active for another reason this year: there are two eclipses in Capricorn—a solar eclipse, January 15, and a lunar eclipse, June 26. The January eclipse is by far the best of the two because of its alignment with Venus, also in Capricorn. That should have you among the favored few at work, and you'll enjoy your job more than ever. Use it for all it's worth and turn in an outstanding performance day after day. Because the eclipse is also aligned with Jupiter and Uranus in Pisces, you'll have an opportunity to earn a nice raise or bonus. Be cautious, though, about romance with a coworker, which could cancel all the potential gains from this eclipse. Each eclipse is in effect for approximately six months.

The June eclipse in Capricorn, however, activates powerful Pluto, which also clashes with Mercury in Cancer. Even in the best work situation you'll need to find a method for stress relief and stick to it. Exercise can be helpful here, and you might also try journaling to sort out your thoughts, feelings, and frustrations. Also maintain confidences and be aware—listen to the grapevine.

July 11 brings the year's third eclipse, this time in Cancer, your solar Twelfth House of self-renewal. In a sense this echoes the Pluto theme, encouraging you to treat yourself well. This solar eclipse is also all about action because it's favorably aligned with Mars in Virgo, your solar Third House. Use the red planet's energy to motivate yourself and to relieve tension. Physical activity can be the best way to quiet your mind before bedtime.

The year ends with a lunar eclipse in Gemini, your solar Eleventh House of friendship and group activities. This is a real plus for holiday get-togethers, and will link some Leos with a new romantic interest. Jupiter and Uranus in Pisces are also activated by this eclipse, which increases the potential for travel in the coming months. This also repeats the year's planetary message of education, and could be just what you need to enroll in school. Take care, though, when you're out and about; you don't need a messy legal situation to deal with as a result of an accident.

Saturn

If you were born between August 20 and 22, you're among the last of your sign to have Saturn contact your Sun from Virgo, your solar Second House. Finances require much care both during this transit and in the months leading up to it. (Saturn will be in Virgo from April 7 to June 20.) Be prepared. Be thrifty. Save more than you spend. Saturn could restrict resources, including income. By establishing this attitude early in the year you'll be in a better position to manage as Saturn concludes its time in Virgo.

But Saturn here also has its upbeat side. You could collect a long overdue debt during this time, or discover that a memento is actually a valuable collectible. Check those boxes of childhood and family treasures.

Saturn in Libra, your solar Third House, will contact your Sun **if your birthday is between July 22 and August 8**. However, **Leos born after July 27** will experience this Saturn transit for only a few weeks between mid-August and year's end. **If you were born July 22–27**, Saturn will contact your Sun between January 1 and April 6, and again in late July and early August. Events that occur earlier in the year will conclude during the second contact. With Saturn in your solar Third House your mind is at its best, but you'll also have a tendency to dwell on things and overanalyze events. Step back and get a fresh perspective when you slip into this mode.

You, more than others born under your sign, should seriously consider the potential benefits of additional education. Even if you can't see the potential benefits now, you'll be in a much stronger position in a few years to capitalize on this through your career. Do it now rather than regret it later.

Uranus

With Uranus in Pisces, your solar Eighth House, you'll want to keep a close eye on finances this year **if your birthday is between August 16 and 22**. Family income could fluctuate and unexpected expenses arise. But you might also gain through a bonus, inheritance, or insurance settlement. The main challenge with this transit is that money matters will be generally unsettled. Your best recourse is to play it safe and build up savings early in the year.

This is even more important **if your birthday is between August 20 and 22** because Saturn in Virgo, your solar Second House, will also contact your Sun. This planetary lineup can impact both your and your partner's finances. An ultraconservative and thrifty financial attitude will help see you through this period, which the universe presents as a learning experience in financial responsibility and stewardship.

If you were born July 22 or 23, Uranus in Aries will contact your Sun from May 27 to August 12, when it briefly visits your solar Ninth House. You may meet someone during this time who has a tremendous, even life-changing, influence on you. For some singles, summer brings a whirlwind romance. Great! But don't rush into a commitment, especially one that involves finances. What seems like a winning liaison ultimately may or may not be in your best interests.

Travel during this time frame can be exciting and enlightening but also prone to delays, and an unexpected trip may be necessary to assist a relative. You also may feel an increasing need to make significant changes in your life in order to redefine yourself and your place in the world. Yet at the same time you'll want to cling to the past as Saturn in Libra and Uranus in Aries pull you in opposing directions. Own the feeling and analyze it as a first step toward blending the past with the future to get the best of both. Aim for realistic, positive, proactive change.

Neptune

If you were born between August 16 and 22, Neptune will contact your Sun from Aquarius, your solar Seventh House of relationships. As great as this transit is for love and romance, it's decidedly chancy for commitment. So resist the temptation to enter into what may not be a lasting tie. Hold off at least until next year. By then you should know if you're in love or in love with love. Financial partnerships in any context are also unwise this year. Avoid debt if at all possible and don't agree to cosign a loan, purchase property with someone who wants the benefit of your credit rating, or loan money to a friend or relative. You're likely to end up on the losing end.

What you can do with this transit is to help others help themselves. Be an inspirational motivator, rather than an enabler, and a mentor who guides others in their search for excellence.

Pluto

Pluto in Capricorn, your solar Sixth House of work and wellness, will contact your Sun **if you were born between July 25 and 28**. Major changes in your workplace are on the horizon, and even if they don't directly affect your job, you'll feel the tension around you. Keep a low profile as much as possible to avoid being pulled into a contentious situation, and be sure to document difficult discussions and events with coworkers and supervisors. Think before you speak, and don't send e-mail in a heated moment.

Find an outlet for stress relief, something you can do on a daily basis such as walking, light reading, or meditation. It will be great for both your mental and physical health, and help ensure a good night's sleep so you're at your best and brightest day after day.

Jupiter-Saturn-Uranus-Pluto

All four of these planets will contact your Sun **if you were born between July 22 and 26**. This unusual alignment, which occurs this summer, will involve your solar Second House (Saturn in Libra), Sixth House (Pluto in Capricorn), and Ninth House (Jupiter and Uranus in Aries). Uranus, as always, is the wild card in this lineup, making it tough to forecast the possibilities. You can be sure, however, that any events will unfold quickly and unexpectedly.

This four-planet lineup could trigger downsizing at your workplace, and even though your job might be safe, this event could be all the motivation you need to return to school—something you should consider in any case. Be very cautious on the road, and catch a ride if you're out socializing. You also can expect travel disruptions and delays, and might need to assist an out-of-town relative. Legal affairs are best avoided if at all possible. And you should defer decisions to a supervisor rather than risk being blamed or becoming involved in a power play on the job. This lineup also has its potential positives: let it motivate you to revamp your general life outlook for a successful future.

Leo/January

Planetary Hotspots
Frustration. Delays. Personal plans put on hold. Expect them all with Mars retrograde in your sign through early March. It's important to identify a daily stress reliever, and to find a way to calm your temper when it rises, as it will, especially the week of the January 30 Full Moon in Leo. Try to see this period instead as a time of preparation, to think about what you want to accomplish this year and next. Focus on personal goals and make concrete plans to achieve them.

Planetary Lightspots
January 17 is the date you've been waiting for. That's when Jupiter enters Pisces, your solar Eighth House, with all the potential to multiply your bank balance in the year ahead. Use the first few weeks of January to outline a money management plan for 2010, so you end the year with more than you have now. Otherwise, it could slip through your fingers.

Relationships
Relationships get a boost from Venus in Aquarius, your solar Seventh House, from the 18th on. You'll attract people into your circle, and possibly a new romantic interest. If you're committed, delight in romantic moments with your mate. Use this transit to cultivate anyone who might be useful in your career within the next few months.

Money and Success
Workplace tension is thick at times this month as Saturn in Libra, your solar Third House, clashes with Pluto in Capricorn, your solar Sixth House. Controlling people make life difficult, and conflict puts others at odds. Give no one cause to question your performance; fulfill responsibilities, and do your best not to get drawn into stressful discussions. Company downsizing could occur, although you may not be directly affected.

Rewarding Days
1, 2, 3, 10, 11, 12, 15, 17, 22, 25

Challenging Days
6, 9, 13, 16, 20, 23, 27, 29, 30, 31

Leo/February

Planetary Hotspots
Relax rather than frustrate yourself. With Mars still retrograde in your sign it's only natural that you still feel like you're spinning your wheels, caught in a personal holding pattern. That won't begin to change until next month, so take advantage of this time to finish the task you began last month by fine-tuning personal plans for 2010. Talk it over midmonth with someone close to you who has an objective perspective.

Planetary Lightspots
The fast pace at work continues through the 9th as Mercury wraps up its tour of Capricorn, your solar Sixth House. Budget your time to manage the many distractions, but also take time to talk with coworkers. You can pick up a lot of information on the grapevine that could prove useful later this year.

Relationships
Your solar Seventh House comes alive under the February 13 New Moon in Aquarius, putting you in touch and in sync with many. This influence is great for couples in love, and some Leos move closer to lifetime commitment. People will inspire you in early and late February, but check facts and ask tough questions if what you hear sounds too good to be true. Midmonth could bring conflict with someone at a distance, possibly a relative.

Money and Success
The February 28 New Moon in Virgo, your solar Eighth House, has financial potential that you could realize as a raise or bonus in recognition of a job well done. Spend a little, but save more. You may need it later in the year for unexpected expenses. If not, you'll have a nice cushion for the future.

Rewarding Days
3, 4, 7, 8, 11, 13, 14, 18, 24, 26

Challenging Days
2, 5, 6, 9, 12, 17, 19, 20, 23

Leo/March

Planetary Hotspots
Avoid travel if you can. Planetary alignments signal an increased chance for delays, cancellations, and lost luggage. Business travel is especially unfavorable—something to keep in mind if you're involved in a presentation or plan to attend a seminar. Also, a legal matter may or may not turn out in your favor. Instead, emphasize the potential positives of these planetary influences and use them to research information and get the knowledge you need.

Planetary Lightspots
Your personal life begins to regain momentum after Mars resumes direct motion on the 10th. As anxious as you'll be to quickly move into high gear, however, a little caution is probably a good idea. Otherwise you could run when you should walk, or make a snap decision it will be difficult to undo.

Relationships
Communication flows under the March 29 Full Moon in Libra, your solar Third House. That's an asset in everything you do, and people will be receptive to you and your requests. But there's still potential for tension and conflict in some relationships, so remind yourself not to jump to conclusions and to think before you speak or write an e-mail.

Money and Success
You could be in the money this month, thanks to the March 15 New Moon in Pisces, your solar Eighth House, and lucky planetary alignments. Again this month, a raise or bonus could come through for you or your partner, and a small windfall is also possible, such as an unexpected win or settlement. Don't spend, however, until you have the cash in hand. It may be as long as six months or more until you see a check.

Rewarding Days
4, 6, 7, 9, 12, 13, 14, 15, 17, 19, 26, 30

Challenging Days
2, 3, 8, 11, 16, 18, 20, 23, 25, 29

Leo/April

Planetary Hotspots
Financial issues are at the forefront as Saturn returns to Virgo and clashes with Uranus in Pisces. This emphasis on your money sectors can trigger a change in income, increased expenses, and difficulties with taxes, an insurance claim, or inheritance. You also could see company benefits reduced. Curb spending, emphasize saving, and make necessary changes using a realistic, practical approach—even if you have no major financial issues.

Planetary Lightspots
The April 14 New Moon in Aries, your solar Ninth House, fuels your desire for knowledge and adventure. Take a trip or browse the library. Take a class or research a subject that's always been of interest. What you learn now will come in handy within a month or two.

Relationships
Your popularity soars with Venus in Taurus, your solar Tenth House, through the 24th. Socialize with coworkers and encourage them to bring their friends along. You can make some great contacts to benefit your career. However, keep a close eye on valuables when you're out and about because money or a treasured possession could be gone in a flash. Also be selective about what you say to whom. Your words are likely to be repeated.

Money and Success
Although career matters are generally positive and upbeat, you'll want to take the time to review facts, figures, projects, and routine work. With Mercury retrograde from the 17th on, you'll ultimately be responsible for any errors that slip through. Pay close attention around the 5th and 25th, when you'll be in a rush, impatient and anxious to move to the next task.

Rewarding Days
2, 3, 8, 9, 10, 13, 14, 17, 26, 27, 30

Challenging Days
4, 5, 7, 12, 15, 19, 22, 23, 25, 28

Leo/May

Planetary Hotspots
With Venus in Gemini, your solar Eleventh House, the accent is on friendship and group activities. Most will be positive, but avoid mixing money with either one (friends and groups), no matter how pressured or generous you feel. The same applies to any organization in which you're involved or hold a leadership position; get the group's okay before you spend, even if that's never been done before.

Planetary Lightspots
Quiet time. You'll cherish it after Venus enters Cancer, your solar Twelfth House, on the 19th. Plan some laid-back evenings at home, engrossed in your favorite leisure-time activity, and try for the same on the weekend. Venus here is ideal for couples in love—pure romance.

Relationships
By the time of the May 27 Full Moon in Sagittarius, your solar Fifth House, you'll be ready for more time on the social circuit. For singles, the lunar energy could spark a new relationship, but also the end of one that isn't working out as hoped. Set aside extra time for your children in the following two weeks; they'll appreciate your devoted attention.

Money and Success
Pending decisions and projects at work come up for review after Mercury in Taurus, your solar Tenth House, resumes direct motion on the 11th. But there's still room for misunderstandings, so clarify rather than assume. Overall, your career gets a nice boost this month, thanks to the May 13 New Moon in Taurus. Use your star status for all it's worth and be the first to step up and volunteer or take charge.

Rewarding Days
1, 5, 6, 10, 11, 14, 15, 24, 27, 28

Challenging Days
2, 4, 9, 13, 16, 18, 22, 23, 26, 29

Leo/June

Planetary Hotspots
Workplace tension rises under the June 26 Full Moon (lunar eclipse) in Capricorn, your solar Sixth House. Someone (or several someones) may have a hidden agenda and attempt to undermine your efforts or manipulate things and play the blame game. Because this Full Moon activates Pluto in Capricorn and Jupiter and Uranus in Aries, your solar Ninth House, unexpected travel is possible, either for business or to assist a relative.

Planetary Lightspots
You sparkle and shine with charisma that goes above and beyond after Venus enters your sign on the 14th. Turn on the charm with everyone you meet, and let your enhanced powers of attraction do their magic. Almost anything you wish for now can be yours.

Relationships
The June 12 New Moon in Gemini energizes your friendship sector, bringing with it the potential for an abundant social life this month. With Mercury in Gemini from the 10th to the 24th, you'll have all the right words for any meet-and-greet. But set financial ground rules before joining friends for a night out. Opt for a less expensive destination and then pay only your fair share. You could meet someone near month's end who turns your life around.

Money and Success
Mars dashes into Virgo, your solar Second House, on the 7th. That's all the energy, initiative, and incentive you need to increase your earnings and your bank balance. Keep that in your thoughts when the fiery planet triggers the urge to splurge. Impulse buys could offset potential gains, and you'll be more than usually tempted to spend on loved ones this month.

Rewarding Days
1, 7, 11, 15, 16, 20, 28, 29, 30

Challenging Days
2, 3, 5, 6, 12, 19, 24, 25, 26

Leo/July

Planetary Hotspots
Saturn returns to Libra and aligns with Uranus in Aries across your solar Third/Ninth House axis. Be very cautious on the road, and always ride with a designated driver if you're out socializing. Don't take the risk. Take a similar approach with difficult neighbors; walk away rather than compound the problem, no matter how tough it is to let it go. Also be prepared to assist a relative, if necessary. On the upside, take the hint if this planetary energy nudges you toward returning to school.

Planetary Lightspots
You'll probably need a relaxing break this month. Take one around the time of the July 11 New Moon (solar eclipse) in Cancer, your solar Twelfth House of self-renewal. Pamper yourself, get a massage, curl up in your favorite chair with a thriller, treat yourself to some comfort food. And don't forget to turn off your phone!

Relationships
The 25th brings the Full Moon in Aquarius, your solar Seventh House. This influence is among the best for close relationships, but you should be a little wary of anyone new who comes into your life. What sounds too good to be true probably is, and an offer or opportunity may not fully materialize despite guarantees.

Money and Success
Finances look much brighter this month, thanks to Venus, which enters Virgo, your solar Second House, on the 10th. It's possible this beneficial influence could trigger a raise, but you're more likely to cash in on bargains if you shop sales for household items and career clothing. Use the last few days of the month with Mercury in Virgo to create or update a budget that includes savings.

Rewarding Days
2, 4, 5, 8, 12, 13, 17, 26, 27, 31

Challenging Days
3, 9, 10, 16, 19, 20, 22, 23, 25, 30

Leo/August

Planetary Hotspots

Recent events reach a climax this month as planetary alignments activate Jupiter, Saturn, Uranus, and Pluto in your solar Third, Sixth, and Ninth Houses. The situation will unfold rapidly, and with Uranus involved, the outcome is more than likely to be unexpected. Downsizing could affect your job, and you may find yourself headed out of town to handle a family issue. Again, you should consider additional education, especially if your job is in some way affected by these transits. Do this and it will begin to pay off next year.

Planetary Lightspots

You're in the limelight! Step out with confidence every day, knowing that the August 9 New Moon casts its beams on your sign. Set personal goals now for the next twelve months. Then put all your determination behind achieving what you set out to do.

Relationships

Despite this month's hotspot that includes your solar Third House, Venus in Libra from the 6th on will help you gain supporters—if you approach them with honest, straight talk. That could be tough to do if you're unwilling to listen and take their advice. Open your mind, put the past in its place, and embrace the future. Self-knowledge is your goal and your key to success.

Money and Success

You'll want to pay bills early this month because Mercury turns retrograde in Virgo, your solar Second House, on the 20th. Take advantage of this time to organize financial records. Four days later, the Full Moon in Pisces lights up your solar Eighth House of joint resources. That can be a plus for your partner's income and investments. But with Uranus returning to Pisces on the 13th, be sure to build up savings to cover any unexpected expenses.

Rewarding Days

1, 8, 9, 14, 17, 22, 23, 31

Challenging Days

4, 6, 12, 16, 18, 19, 20, 27

Leo/September

Planetary Hotspots
Although this summer's major planetary alignment is waning, you'll still have some related issues to deal with in early September when Jupiter in Aries clashes with Pluto in Capricorn. This duo could trigger more travel or the conclusion of legal matters, as well as your return to school (go for it!). You might also decide to delve into your roots to learn about the past in order embrace the future by learning more about your family history.

Planetary Lightspots
Cross your fingers. The lucky Jupiter-Uranus merger in Pisces, your solar Eighth House, could trigger a unexpected windfall. This lucky influence could also spark a chance encounter with someone who helps make your dreams come true.

Relationships
Family relationships are at their best after the 7th, when Venus is in Scorpio, your solar Fourth House. With luck, you might even get family members to pitch in and help with domestic projects once Mars enters the same sign on the 14th. Get everyone organized and get busy on anything from the yard to winterizing your home to redoing a room or two. Start now and you can finish it all by the end of October.

Money and Success
September brings a double financial emphasis. The New Moon in Virgo, your solar Second House, on the 18th also has the potential to boost your bank account, especially after Mercury resumes direct motion on the 12th. But this transit is about more than money. It's also about what you value. That could become clear in an insightful moment that helps you see the best in the people who are closest to you.

Rewarding Days
2, 5, 6, 8, 10, 14, 19, 20, 24, 28

Challenging Days
3, 9, 11, 15, 16, 18, 21, 23, 25, 26, 30

Leo/October

Planetary Hotspots
The stresses and strains of the past few months ease for the most part and you're ready to move forward. Domestic plans and projects, however, may not go exactly as planned. Think of this as a message from the Universe to go with the flow and let things evolve as they will. Ultimately, you'll get the desired results if you get advice from a pro rather than a close friend or family member. Take advice from either one and you could end up with extra expenses.

Planetary Lightspots
Your solar Third and Ninth Houses are active this month, with the October 7 New Moon in Libra and the Full Moon in Aries on the 22nd. That's all the reason you need to plan a vacation or dash off for a long weekend. Delays are possible early in the month, however.

Relationships
Family relationships will be somewhat strained this month after Venus in Scorpio, your solar Fourth House, turns retrograde on the 8th. The influence will be more subtle than overt, but you'll miss the usual warmth and much will be left unspoken. Communication will open up somewhat after Mercury enters Scorpio on the 20th. See this as just another cycle that will soon pass, and welcome the extra time with your own company.

Money and Success
Your work life perks along this month with only the usual hassles. But question what you hear if someone comes on too strong and tries to push what is probably a personal agenda the third week of October. At the same time you'll get the inspiration you need to outdistance the competition. Put forth the effort and rewards will be yours before year's end.

Rewarding Days
7, 8, 10, 11, 15, 16, 18, 21, 26, 29

Challenging Days
2, 3, 5, 6, 13, 20, 22, 23, 25, 27, 28

Leo/November

Planetary Hotspots
Life settles down to a mostly predictable routine this month with small hurdles here and there. Money again flows your way after Jupiter turns direct in Pisces, your solar Eighth House, on the 18th, but that will encourage you to spend more than you should on family, fun, and romance. Scale back and live within your budget.

Planetary Lightspots
Home life is at its best under the November 6 New Moon in Scorpio, your solar Fourth House, and you'll feel all is right in the world when surrounded by those you love. Carry it a step further. Kick off the holidays with a Thanksgiving celebration at your place and invite friends to join you and yours.

Relationships
November 18 is significant. That's the date Venus resumes direct motion in Libra, your solar Third House. Besides being a real plus for relationships in general, you'll have more contact with neighbors and your extended family. But not everyone you encounter this month will be entirely truthful. Be wary and listen to your intuition if anyone tries to gloss over facts. Enthusiasm goes only so far. Also take a dim view if a family member or close friend asks for a loan; you're unlikely to be repaid.

Money and Success
You'll be stretched thin the week of the November 21 Full Moon in Taurus, your solar Tenth House, trying to keep up with career and family responsibilities. Yet the push will be worth it because December will be an even busier work month. If you plan to shop for a necessary big-ticket item, try to hold off until Venus turns direct, and then be sure to confirm the warranty terms and return policy.

Rewarding Days
4, 5, 7, 8, 12, 13, 16, 20, 22, 26

Challenging Days
1, 3, 9, 10, 15, 17, 23, 24, 27, 30

Leo/December

Planetary Hotspots

Mercury turns retrograde in Capricorn, your solar Sixth House, on the 10th. That can trigger problems on the job when decisions are delayed or reversed. And even though you'll be short on time, set aside the necessary hours to review, revise, and check for errors, which retrograde Mercury is noted for. Also confirm dates and times of appointments and meetings. Do the same for social events after Mercury retreats into Sagittarius, your solar Fifth House, on the 18th.

Planetary Lightspots

Take advantage of Venus in Scorpio, your solar Fourth House, all month. Decorate for the holidays, host a few get-togethers, and plan a special family celebration. Most of all, enjoy the peace and serenity of the comforts of home after another long day.

Relationships

The December 5 New Moon in Sagittarius and the Full Moon (lunar eclipse) in Gemini on the 21st energize your solar Fifth and Eleventh Houses—an ideal match for your social life. It's even possible you'll have multiple invitations for some dates. Some Leos will launch a new romance this month, but keep your options open if you're among them. You could have a change of heart in January.

Money and Success

Get set for fast-paced days at work from the 7th on, the date Mars enters Capricorn. There could be a nice year-end bonus to reward your efforts, but you can also expect a higher stress level. Difficult, obsessive people may try to block progress and it will be tough to reason with them. Try to stay focused and calm, and think about quiet evenings at home.

Rewarding Days

1, 4, 5, 8, 10, 13, 19, 23, 29, 31

Challenging Days

7, 9, 11, 14, 15 17, 20, 21, 27, 28

Leo Action Table

These dates reflect the best—but not the only—times for success and ease in these activities, according to your Sun sign.

	JAN	FEB	MAR	APR	MAY	JUN	JUL	AUG	SEP	OCT	NOV	DEC
Move	8, 9		4, 5		25, 26				16–30			2–4
Start a class						19, 20		9–18		4–15		
Join a club	25, 16		22			10–19						
Ask for a raise		1						11	7			
Look for work	2–19				2–4				16, 17			
Get pro advice	21, 22	12, 17										7, 8
Get a loan		19–28	2–6									
See a doctor	18, 19	24	9						16, 17	1		
Start a diet		9, 10	9, 10	5, 6	30, 31	26, 27						
End relationship			11, 12	7, 8	5, 6	1, 2, 29, 30	26, 27					
Buy clothes	11	7	3			24		17				5
Get a makeover			25, 26			15, 16		9, 10			26, 27	
New romance						24, 25		17			26, 27	5, 6
Vacation			18–31									

The Virgin
August 22 to September 22
♍

Element: Earth
Quality: Mutable
Polarity: Yin/feminine
Planetary Ruler: Mercury
Meditation: I can allow time for myself
Gemstone: Sapphire
Power Stones: Peridot, amazonite, rhodochrosite
Key Phrase: I analyze
Glyph: Greek symbol for containment

Anatomy: Abdomen, gall bladder, intestines
Color: Taupe, gray, navy blue
Animal: Domesticated animals
Myths/Legends: Demeter, Astraea, Hygeia
House: Sixth
Opposite Sign: Pisces
Flower: Pansy
Key Word: Discriminating

Your Strengths and Challenges

Virgos have a strong work ethic, and concrete results motivate you to strive for more. Your practical approach is task-oriented, and checking items off a to-do list gives you a great sense of accomplishment. Methodical and detailed-oriented, your excellent analytical skills and Mercury, your ruling planet, help you zero in on the task at hand.

But with so much attention to detail and analysis, you can get mired in the small stuff and miss the big picture. Each is important and you have the unique ability to get the best of both. All it takes is a different perspective. Step back and view each task as a part of a bigger whole and see where it fits into the larger scheme of things. You may not realize you do this every day, but you naturally use your keen powers of observation and eye for detail to connect seemingly unconnected information. A glance, a casual comment, body language, bits and pieces—before long it all comes together in a flash of insight.

Many Virgos are shy in their younger years, but develop self-confidence with life experience and by pushing themselves to develop their talents and people skills. Virgos also have a reputation for neatness. Some are, some aren't. Either way, you have a knack for keeping track of everything in a multitude of piles, what others might describe as an organized mess.

Your Relationships

When it comes to people, your strength is being with those you know well. That's your comfort zone; meet-and-greet is not. Even if you know a lot of people, your close friends are few and you treat them like family. In a sense, they are part of your extended family, and you enjoy doing things with and for them, providing for their every need when entertaining at home. Your closest friendships are for a lifetime.

Virgos are as discriminating in love as they are about all else in life. That can make it tough to connect with a new romantic interest, especially if you focus only on your perceived "type" and never take a chance. You might also have a long-term relationship that goes on and on without commitment. Until you feel the zing, that is! Your ideal match could be a dreamy Pisces, your opposite sign, and lasting love is possible with one of the other earth signs, Taurus and

Capricorn. A Cancer or Scorpio might appeal to you, but Gemini and Sagittarius probably won't be the best match.

You're fond of your family even if you live far away from them, and learning was probably emphasized during your childhood years. Your parents may have encouraged you to explore your interests and the world. Most Virgos aren't especially motivated to have children, and usually have small families. As a parent you're supportive but also have high expectations. Go easy and remember that praise and constructive comments foster success.

Your Career and Money

Your career is perpetually subject to change, and this area of your life is constantly evolving as you add new interests. You have a knack for pulling these varied interests together into a whole, and each major career event sets things in motion for the next. In your day-to-day work you need the freedom to create your own structure and prioritize responsibilities. A high level of autonomy is a must, and a hovering boss is almost guaranteed to make you look elsewhere. Your dream job would be one that lets you set your own hours or telecommute.

Your finances are generally in positive territory because of your earning power and conservative nature. You're a saver and can do well with long-term investments. Most Virgos dislike using credit and prefer to pay cash whenever possible. Usually thrifty, you'll nevertheless splurge at times, especially on loved ones—sometimes too much so. Treat yourself the same; occasional self-indulgence is a very good thing!

Your Lighter Side

Your mind is packed with information—everything from directions to gossip to facts and figures. Practical know-how comes from reading, listening, and asking questions. The rest comes from people who willingly share confidences because they know you won't repeat a word. That, combined with your attention to detail, gives you a sixth sense that's among the best.

Affirmation for the Year

People are my source of knowledge and my lucky charm.

The Year Ahead for Virgo

As a hard-working Virgo you'll undoubtedly be somewhat sorry to see bountiful Jupiter depart Aquarius, your job sector, January 17. It moves on to Pisces, where it spends most of 2010, except for a few months in Aries.

Get set to welcome many new people into your life as Jupiter advances in Pisces, your solar Seventh House. Some will become friends, others will be supportive coworkers, and still others will appear for a specific purpose. This transit is one of the best for Virgos searching for love, and you could meet the man or woman of your dreams and get lost in the world of romance. Family members will also have a more prominent role in your life, and you'll develop stronger ties with some. At least one of them will bring you luck. In general, you have a special role this year: encouraging others to do and be their best. Spread optimism and enthusiasm!

Summer could be a lucrative season with beneficial Jupiter in Aries, your solar Eighth House, June 5–September 7. Jupiter expands whatever it touches, so cross your fingers for a sizeable raise or bonus or a nice windfall. Test your luck on the lottery or a sweepstakes entry. Be aware, though, that when Jupiter is involved with money matters it can just as easily expand debt and promote big spending. Save and invest with caution, and look forward to 2011, when you'll get the full benefit of Jupiter in Aries.

Just when you thought you'd seen the last of Saturn, it does a rerun, returning to your sign April 7 for a brief and final stay through July 20. Saturn is in Libra the rest of the year.

Think back to 2007, when Saturn began its Virgo transit and mentally review the ensuing years and events that occurred. Among them is at least one that you can call unfinished business. Thank Saturn for this extra opportunity to bring things to conclusion and complete what you began in the past few years.

Saturn is in Libra, your solar Second House, the rest of the year, where it will remain the next few years. Saturn here has a reputation for restricting income. Although that can be true, more people experience this transit as "Depression mentality"—thriftiness taken to the extreme. That's the result of a fear of financial insecurity. Aim for the middle ground—savings and thrifty, value-based spending.

Also develop a budget to pay off any accumulated debt as quickly as possible. Get in the habit of using cash or debit rather than credit.

Saturn's Second House transit is about more than money, however. It's also about possessions and what you most value. On a practical level, the time has arrived to clean out closets and to generally get your space in order. Take the best to a consignment shop and donate the rest. While you're at it, keep an eye out for what could be valuable collectibles. Ask yourself what you value and give this question serious thought. Things? People? Your skills and talents? Family? Friends? Yourself? It may take much of Saturn's time in the Second House for you to fully define what truly matters in your life. As with everything, Saturn is slow, so you'll gradually see new attitudes emerge from within during the next few years.

Uranus continues its multi-year transit of Pisces, with a brief foray into Aries, May 27–August 12. With Uranus, planet of change, in Pisces, your solar Seventh House of relationships, you've undoubtedly seen people come and go in your life in recent years. That pattern not only continues but accelerates in 2010 because of Jupiter, also in Pisces.

The Jupiter-Uranus merger in Pisces bodes well for love at first sight. Enjoy every romantic moment if you're among them, but it might be wise to postpone commitment until next year. What begins in a flash could end in a flash. Take note of chance encounters this year. These people, who seem to appear out of nowhere, will arrive for a reason just when you need them. Don't be surprised, however, if you consciously end a relationship or two, or if someone in your life does the same. What worked in the past may no longer be a positive. But think carefully before you act. Uranus endings are almost always final. Some coworker relationships are also likely to change. Be cautious here. Someone who befriends you on the job could be more foe than friend.

As much as Jupiter in Aries, your solar Eighth House, can fatten your bank account, Uranus can do the same—or exactly the opposite. This volatile and quirky planet is known for sudden ups and downs when it influences money matters, which can be anything from a million-dollar win to expensive repairs. Avoid risky investments and be sure not to overlook a tax payment. In fact, you'd be wise to double-check that bills are paid and payments received during the months that Uranus is in Aries.

Your work life will go through inspirational periods as well as disappointing ones just as it has since 1995, when Neptune entered Aquarius, your solar Sixth House. This mystical planet of illusion and confusion can make it tough even for painstaking Virgos to focus on the task at hand and to grasp all the details.

There's an easy solution: get creative! A different perspective can help you see things as they are and what others miss. You can accomplish this by switching tasks, taking a walk, or looking at things in reverse order. Take this a step further and offer your creative ideas in meetings and talks. While they might first be perceived as off-base, people will soon appreciate your vision and imagination. Use both in problem-solving. One cautionary note: keep personal business to yourself on the job.

Pluto in Capricorn, your solar Fifth House, has a message similar to Neptune: develop your creative energy and transform your life. True, that's something of an oversimplification of what you'll experience during Pluto's long transit of Capricorn (until 2024). But not nearly as much as it might first seem.

In a sense Pluto in Capricorn encourages you to redefine yourself, to move outside your comfort zone into areas that are intangible but that have tangible results. Look within and find and begin to express your practical creative self. An easy way to get started is to learn a new hobby such as furniture refinishing, specialized gardening (breeding roses or growing orchids, for example), home improvement skills, or writing. The maximum benefit over time will be increased confidence and self-knowledge.

You'll also be more involved in your children's lives in the coming years, and will experience tremendous personal growth from parenting. In turn your children will be the catalyst that motivates your personal change.

What This Year's Eclipses Mean for You

The year's two eclipses in Capricorn, one solar (January 15) and one lunar (June 26), reinforce the Pluto message: take charge, get motivated, and express yourself! Both eclipses are excellent for delving into this hidden side of yourself and letting your inner voice emerge. Each eclipse is active for approximately six months.

Both these eclipses also put romance in the forecast. They're among the best whether you're searching for love or half of a couple.

The January eclipse is particularly favorable because it's beautifully aligned with Venus, also in Capricorn, and Uranus in Pisces, your solar Seventh House. So get out and socialize at every opportunity to increase the odds of meeting your soul mate.

The June 26 lunar eclipse also favors romance, but because it's aligned with Pluto a relationship could become decidedly unhealthy and all-consuming. This eclipse is also aligned with Jupiter and Uranus in Aries, so don't let anyone, especially a romantic interest, convince you to part with your hard-earned money.

Either of these eclipses could trigger an addition to the family for some Virgos. Proud parents will celebrate their children's accomplishments but can also expect increased expenses for their activities. Also resist any temptation to program every hour of your children's lives. They'll do far better when given age-appropriate freedoms.

July 11 brings the year's third eclipse (solar) in Cancer, your solar Eleventh House of friends and group activities. Like the Capricorn eclipses, this one has potential to ramp up your social life along with the opportunity to widen your circle of friendship and networking contacts. You can also do this by getting involved in a professional organization or a good cause that would welcome your skills and talents. The Eleventh House also relates to goals and objectives, what you want and wish for. So give some thought to what you want to achieve the second half of 2010, make an action plan, and get started. With Mars in your sign at the time of the eclipse you'll have the initiative and incentive to make it happen.

The year ends with the December 21 lunar eclipse in Gemini, your solar Tenth House of career. You'll carry this energy with you into 2011, when your ambitions will soar. But focusing on an exact path could take a little time because this eclipse is closely aligned with retrograde Mercury, your ruling planet. The eclipse also clashes with Jupiter and Uranus in Pisces, your solar Seventh House, so look to someone close to help you define your goals and choose the best career path.

Saturn

If you were born between September 20 and 22, Saturn in Virgo will make its final contact with your Sun from April 7 to July 20. Recall the events that occurred last October and what you learned from the experience. This is the significance of Saturn's transit to your Sun, and what you should resolve or conclude before the serious planet departs your sign. Keep in mind that Saturn's contact with your Sun is karmic; it brings you what you deserve, whether it's a personal accomplishment or a letdown. Saturn nearly always rewards hard work, responsibility, and playing by the rules. Aim high and launch yourself in a new direction.

You'll want to plan ahead for downtime during Saturn's transit, which can indicate diminished energy. Sleep and relaxation can help, as can moderate exercise. Then you'll have the mind and body power you need to get the best of Saturn's determination to fulfill an increasing desire to organize and re-order your life. At the same time, go easy on yourself and be satisfied with what you accomplish rather than what you don't. The same applies to any life regrets that surface now; look to the future, not the past, to what might be, not what might have been.

Saturn in Libra will contact your Sun **if you were born between August 22 and September 8**. You'll have the greatest Saturn influence **if your birthday is August 22–27**. (Otherwise, you can expect a week or two of Saturn between mid-August and year's end.) You'll want to keep close tabs on spending and do all you can to increase savings as Saturn transits Libra, your solar Second House, January 1 to April 6, and again from July 20 to mid-August.

Actions taken and decisions made in the earlier period will come up for review during Saturn's second Libra transit of the year, so getting your finances in order should be a priority the first three months of 2010. A conservative mindset will benefit your bottom line, and thrifty shopping can not only save money but net you some fantastic deals on necessities.

Saturn's Second House lesson, above all, is: learn to manage personal resources, both when money is tight and when things are status quo. Adopt the new habits and attitudes that will be rewarded for many years to come and that will culminate in about fourteen years when Saturn enters your solar Eighth House of joint resources.

Uranus
If you were born between September 15 and 22, Uranus in Pisces will contact your Sun between January 1 and May 26, and again

between August 13 and December 31. Relationships will be in flux with Uranus in your solar Seventh House, and you'll part ways with some people. New people will appear as well, some of whom could be invaluable contacts in the future.

But as much as this transit is about other people it's even more about you and your changing needs. This is especially true **if your birthday is between September 20 and 22**, because Saturn in Virgo will form an exact alignment with Uranus (April 26) across your solar First/Seventh House axis. Blending these two energies is a challenge; Saturn prefers the past while Uranus looks to the future. So you'll feel the push-pull of keeping things as they are versus moving on. Ultimately, the decision about any relationship will center on compromise, whether you can find a middle ground or whether it's best to move on in order to fully pursue your dreams.

If you were born August 22 or 23, Uranus in Aries will contact your Sun between May 27 and August 12. Your bank balance is likely to fluctuate, perhaps dramatically, with the planet of change in your solar Eighth House. That's all the more reason to stash cash for unexpected expenses and a potential decline in family income. You'll be more prone to these sudden shifts because Saturn in Libra, your solar Second House, will be aligned with Uranus. But it's also possible you could gain from a lucky win, return on an investment, or an unexpected inheritance. Be cautious, though. Even what looks like a sure deal can backfire when Uranus is involved.

The Eighth House also governs insurance, so be sure your property is well-covered and all premiums are up to date. This is not the time to take chances with anything financial, including taking on debt that's based on anticipated future income or loaning money to a friend or relative.

Neptune

If you were born between September 16 and 22, Neptune will contact your Sun from Aquarius, your solar Sixth House of daily work and wellness. Confusion and chaos are possible in the work place with this transit, and it might be tough to get a handle on exactly what's happening both with your job and your company, especially in May and November. The same time frames could have you feeling disillusioned about your prospects for the future and whether you're on the best path for you. Finding the answer, however, will be difficult if not impossible this year because your thoughts and ideas will shift in sync with illusory Neptune. You'll also want to be very cautious about sharing personal information with coworkers, who could use the information to their advantage. Double-check your work; mistakes are easily made now.

Like every planet, Neptune also has its positive side. In the Sixth House it can inspire you to go above and beyond, to put forth the extra effort that's both personally and professionally rewarding. Creative thinking and approaches can contribute to this success, and you might find yourself thrust into the limelight as a result. You might also want to consider volunteering your time for a charitable organization, but resist the urge to provide for friends, coworkers, or family members. Inspire them instead.

Pluto

If you were born between August 26 and 28, Pluto in Capricorn, your solar Fifth House, will favorably contact your Sun. This may be just the year you've been waiting for, the one in which you'll grow into yourself, confident and empowered. Pluto's dynamic energy can help you accomplish this whether you want to get in shape, embrace a bold new direction, or achieve what you never thought possible. This powerful planet's willpower and determination is yours for the asking. Use it to the max!

A new child will be a life-changing event for some Virgos, and if you're a parent you can do much to support and encourage your children, guiding them on the right path through life. Listen closely to what they say. Their innocent comments could trigger insights into your own life and how to become the best you can be.

You'll also be among those born under your sign who can easily tap into Fifth House creativity. Find a way to express your individuality because this too will contribute to the person you are becoming. And be sure to give yourself permission to make mistakes; no one learns anything new overnight, and perfection is a dream, not reality.

Jupiter-Saturn-Uranus-Pluto

If you were born between August 22 and 26, this unusual four-planet alignment will contact your Sun this summer. There are many different scenarios that could unfold, not the least of which is a sizeable financial gain. But financial loss is just as possible so don't gamble or invest whatever you can't afford to lose. And, remember, it takes only one lottery ticket to be the big winner. You (or your mate) might have a significant unexpected expense, receive a nice inheritance or, unfortunately, be caught up in company downsizing. Entrepreneurial ventures are risky at best.

Because the Fifth House is involved in this configuration, child-related expenses could be steep, possibly because you're the parent of an emerging athletic or academic star, or one who could make it as an actor. Carefully weigh the risk versus reward; this may not pay off until next year, if at all.

Check all insurance policies to be sure you're adequately covered, and don't take anyone's word regarding money matters on faith. Do the same if for some reason you must sign a contract or negotiate a settlement; read all the fine print.

If you're a parent and you do net a nice windfall, consider starting a college savings account for your children. You'll be glad you did when the time comes. And set aside some for you in a retirement account.

Virgo/January

Planetary Hotspots
Finances require caution as Saturn in Libra, your solar Second House, squares off with Pluto in Capricorn, your solar Fifth House. Expenses could rise and income decrease under this contact. Most of all, don't risk money on speculative ventures, no matter how promising they look around the time of the January 15 New Moon (solar eclipse) in Capricorn. Save, rather than indulging loved ones.

Planetary Lightspots
The January 30 Full Moon in Leo, your solar Twelfth House of self-renewal, reminds you that relaxation is good even for Virgos. That's particularly important now with Mars retrograde in Leo. Get lost in a page-turner every evening or release stress with a workout at the gym to ensure a good night's sleep and high-energy days.

Relationships
Lucky Jupiter enters Pisces, your solar Seventh House, January 17, marking this as a year with great potential for love, romance, close friendships, and helpful people. Some Virgos will meet a soul mate, while others will form a strong and lucrative business relationship. If you need a roommate to share expenses, start searching after Mercury turns direct on the 15th, and look toward a friend or coworker rather than a romantic interest.

Money and Success
Your work life is both satisfying and upbeat after Venus enters Aquarius, your solar Sixth House, on the 18th. Try to plan ahead because you'll be scrambling at month's end to complete your to-do list, possibly because someone else fails to complete a project. You might even earn some extra money for going above and beyond. Be cautious, though, about e-mail; reread before you send.

Rewarding Days
1, 3, 5, 14, 15, 17, 18, 19, 25, 28

Challenging Days
4, 6, 8, 11, 13, 16, 20, 23, 27, 30

Virgo/February

Planetary Hotspots
February is an easygoing month for the most part. But with Mars still retrograde in Leo, your solar Twelfth House, you can expect frustration and a rising temper at times. Put this energy to good use around the house by getting a jump start on spring cleaning. Organize closets, drawers, and financial records, and make a little extra money by taking discards to a consignment shop.

Planetary Lightspots
Give in when the February 28 Full Moon in your sign encourages you to do something terrific for yourself. Enjoy a spa day, take time off, read a thriller, and spend quality hours with your favorite people. A creative project is great for an ego boost, and you'll gain new insights by listening closely to your children's simple but profound words of wisdom.

Relationships
Relationships continue to be upbeat this month, thanks to Venus in Pisces, your solar Seventh House, from the 11th on. Take a chance on someone new, or romance your partner with a Valentine's Day celebration. Someone you meet midmonth, possibly in a chance encounter, could be the lucky charm you've been waiting for. Meet and greet. Network. Trade favors.

Money and Success
The February 13 New Moon in Aquarius, your solar Sixth House, spotlights your work life and infuses it with positive energy. But keep your wits about you. People will have a tendency to tell you what you want to hear rather than reveal the full truth. Balance idealism with reality, emphasize teamwork, and use innovation in problem solving and projects.

Rewarding Days
1, 4, 6, 10, 11, 13, 14, 15, 16

Challenging Days
2, 5, 9, 12, 17, 19, 20, 23

Virgo/March

Planetary Hotspots
Investments, loans, company benefits, and other Eighth House matters are this month's hotspot with several planets transiting Aries. It could be difficult to get credit now, and you should also check your and the family's credit reports for errors. Insurance coverage may be reduced or premiums rise, and it's possible your partner could see a decrease in income. Use the planetary energy to review finances and to develop a thrifty budget and a debt-reduction plan.

Planetary Lightspots
The pace begins to return to normal as Mars in Leo resumes direct motion on the 10th. Although it will be a few weeks before things are really perking along, you can again move forward with confidence. Tap into the Martian energy and take proactive action regarding finances as well as communication that you've let slide.

Relationships
Relationships get all the benefit of the Sun, Mercury, Venus, and the March 15 New Moon in Pisces, your solar Seventh House. People will respond to requests, and you'll be in sync with just about everyone from friends to family to coworkers. Talks go well and you'll hear uplifting news from someone close to you in early March, and a surprise announcement midmonth.

Money and Success
Even though money matters are strained this month, there's also an upside—the March 29 Full Moon in Libra, your solar Second House of personal resources. The lunar influence could trigger a raise in the following two weeks, or simply help you work through Eighth House challenges. Snap up any opportunity to earn extra money, which could come through your job or a friend or family member.

Rewarding Days
1, 4, 5, 9, 10, 12, 13, 14, 15, 19, 27, 28

Challenging Days
2, 3, 8, 11, 16, 20, 23, 25, 29

Virgo/April

Planetary Hotspots
Saturn returns to Virgo this month and aligns with Uranus in Pisces across your solar First-Seventh House axis. Strained relationships accompany this lineup when you feel the push-pull of the past and the future, the status quo and the unknown. A close tie may end under this alignment, reflecting changing interests and goals. But you also could find a compromise that works for both of you. Postponing the decision until late next month might be wise because Mercury, your ruler, turns retrograde on the 17th.

Planetary Lightspots
With Venus in Taurus, your solar Ninth House, through the 24th, it's time to think about a spring getaway. Browse the Internet for possible destinations and make reservations before Mercury turns retrograde. You'll be glad you did when May brings a desire to get away from it all.

Relationships
Chances are, everyone will want your attention during the two weeks following the April 28 Full Moon in Scorpio, your solar Third House of communication. You'll have plenty of contact yourself with neighbors, relatives, and people at a distance. Most will be positive, some will be confusing or aggravating, and others will share a confidence. Take it all in and put your intuition to work to complete the picture.

Money and Success
Family finances are in the black, thanks to the April 14 New Moon in Aries, your solar Eighth House. Take advantage of the current trend to save money. This summer could bring unexpected expenses, so you'll want to have a cushion. A raise is possible this month, or you might have an opportunity to earn some extra money.

Rewarding Days
1, 5, 6, 9, 10, 11, 16, 20, 27, 29

Challenging Days
4, 7, 12, 15, 19, 22, 23, 25, 28

Virgo/May

Planetary Hotspots
Last month's relationship issues continue to need attention as Saturn in Virgo aligns with Jupiter in Pisces, your solar Seventh House. This alignment, however, signals a definite turning point with an even greater chance for compromise. Then again, you may decide you'd rather go it alone than maintain the status quo, that your future looks brighter if you move on. Keep in mind, though, that what you want now you may regret in the not-too-distant future.

Planetary Lightspots
New places and spaces attract your interest under the May 13 New Moon in Taurus, your solar Ninth House. Take off for a week or weekend for a relaxing change of scenery with family, partner, or friends. Single? You could connect with a potential love interest while traveling on business or attending a reunion.

Relationships
You're drawn to home and family as the May 27 Full Moon in Sagittarius shines brightly on your solar Fourth House. Consider devoting a few days to working around the house. Chances are, family members will be eager to pitch in. Also take time to connect with out-of-town relatives, who will be delighted to hear from you.

Money and Success
You're on top of the career world, thanks to Venus in Gemini, your solar Tenth House, through the 18th. Even better, you'll have plenty of supporters and few detractors. But there's a catch. Be sure to deliver everything as promised. Do that and your star will rise even higher. Also take advantage of any opportunity to meet or get better acquainted with decision-makers. These ties could come in handy this summer.

Rewarding Days
3, 7, 8, 12, 17, 20, 21, 25, 30, 31

Challenging Days
2, 4, 5, 9, 16, 18, 22, 23, 26, 29

Virgo/June

Planetary Hotspots

The June 12 New Moon in Gemini, your solar Tenth House, is both a hotspot and a lightspot. This fresh energy enhances career success, but difficult people could limit that potential—as could you. Curb your impatience and frustration, see the big picture rather than only the details, and follow instructions even if you think you know better—that may be true, but let decision-makers prevail.

Planetary Lightspots

High energy is yours beginning June 7, the date Mars dashes into your sign. Expect to be on the go as you haven't been in months, but slow down enough to prevent a mishap. Also try for daily exercise (at least a quick lunchtime walk) to relieve stress, and calm your mind before bedtime to ensure a good night's sleep.

Relationships

Your social life benefits from the June 26 Full Moon (lunar eclipse) in Capricorn, your solar Fifth House. But planetary alignments could trigger conflict with a friend or love interest, so much so that you might cut ties and move on. Money may be the issue, or you could suddenly discover that your beliefs are at odds. If you're a parent, your children will want more of your time and attention now. Also get to know their friends and their friends' parents.

Money and Success

You or your partner could luck into a nice windfall this month, thanks to a Jupiter-Uranus merger in Aries, your solar Eighth House. Take a chance on the lottery in early May, and watch for sales on big-ticket items. You could find a sensational bargain. But don't take chances with investments or retirement funds because a sure gain could quickly become a sure loss.

Rewarding Days
1, 4, 8, 9, 13, 14, 18, 22, 27

Challenging Days
3, 10, 12, 19, 23, 24, 25

Virgo/July

Planetary Hotspots
Saturn returns to Libra on the 21st, and several days later aligns with Uranus in Aries across your solar Second/Eighth House axis. Expect financial stress and strain with this lineup that could include decreased income, increased expenses, and reduced benefits. Think thrifty and do all you can to minimize spending because this influence will last into the fall months. If a major purchase is unavoidable, consider an extended warranty.

Planetary Lightspots
You'll sparkle and shine from the 10th on as Venus transits your sign. This once-a-year occurrence is one of the best for your powers of attraction, and you'll be able to charm almost everyone into seeing things from your perspective. Use your intuition and a soft sell to increase the odds of success.

Relationships
July is friendship month, with the New Moon in Cancer, your Eleventh House, on the 11th. Even better, favorable planetary alignments will have you in touch with many pals along with the potential for an active social life. Make a point to get together with long-time friends and to reconnect with people you haven't seen in several years.

Money and Success
The July 25 Full Moon in Aquarius is one designed to please hard-working Virgos. Placed in your solar Sixth House, the lunar energy puts you in the job spotlight. But you also could be on the hot seat with a deadline as well as increased workload. If you're job-hunting, the Full Moon could trigger an offer; think carefully and weigh all the pros and cons even if the money is less than expected.

Rewarding Days
1, 2, 6, 7, 11, 14, 15, 19, 24

Challenging Days
3, 8, 9, 10, 16, 20, 22, 23, 30

Virgo/August

Planetary Hotspots
Finances continue to be the hotspot as planets in your solar Second and Eighth Houses clash with Pluto in Capricorn. The downward trend could be due to reduced income and also increased expenses related to your children or as a result of unwise investments. Check your (and the family's) credit reports, and do all you can to avoid added debt. It may be difficult now to get a loan and you should take action with lenders if you see a potential problem. There could also be difficulty with an inheritance or insurance settlement.

Planetary Lightspots
The August 9 New Moon in Leo, your solar Twelfth House, encourages you to take a break, even if only for a day or two. Kick back at home during a long weekend, tune out the world, relax, and re-center yourself with time alone. You'll be glad for the time out when the pace picks up later this month.

Relationships
Uranus returns to Pisces, your solar Seventh House, on the 13th. And while that could prompt change in some close relationships in the coming months, the Pisces Full Moon on the 24th highlights an upbeat period. Plan ahead to spend extra time with your favorite people as summer fades into fall. Share your thoughts and feelings.

Money and Success
Mercury, which rules your sign and also Gemini, your career sector, turns retrograde in Virgo on the 20th. Expect at least some frustration when plans and projects are delayed, and be sure to put your eye for detail to good use at work. Also be cautious with e-mail; re-read before you send.

Rewarding Days
2, 3, 11, 15, 16, 21, 23, 29, 30, 31

Challenging Days
4, 5, 6, 7, 10, 12, 18, 19, 25, 27, 28

Virgo/September

Planetary Hotspots
Although this summer's financial planetary emphasis is still active, it does begin to lessen. It might not feel that way, however, around the time of the Aries Full Moon on the 23rd, which activates Pluto in Capricorn and Saturn in Libra. But this lunar energy also signals a fresh start that comes with the incentive and initiative to take charge of money matters. Develop a plan, put it in writing, and stick to it.

Planetary Lightspots
The September 12 New Moon in Virgo signals the start of your solar year. Better yet, Mercury, your ruling planet, resumes direct motion in your sign on the 12th. This is a terrific combination for personal plans and new directions, as well as relationships. Step out with confidence and set a high bar of personal achievement.

Relationships
Jupiter returns to Pisces, your solar Seventh House, on the 8th, where it joins forces with Uranus on the 18th. This fortunate planetary duo could have you in the right place at the right time to make a lucky contact. For some Virgos it will be a career connection, and others will take a step toward commitment or fall in love at first sight. Family ties will be equally uplifting and rewarding.

Money and Success
Although the Full Moon can limit resources, early September may bring an opportunity to counteract that trend in advance. With Venus in Libra, your solar Second House, through the 7th, and Mars in the same sign through the 13th, you could see your bank balance rise. But this duo also can encourage impulse buys, so remember your budget, be thrifty, and shop sales.

Rewarding Days
2, 8, 10, 12, 16, 17, 20, 21, 22

Challenging Days
1, 3, 9, 13, 15, 23, 25, 29, 30

Virgo/October

Planetary Hotspots
Although life begins to settle down this month, you'll have to contend with Venus, which turns retrograde in Scorpio, on the 8th. This influence can make it tough to get the information you need because people will be generally less helpful and somewhat aloof, especially relatives, neighbors, and those at a distance. Settle in with a book if you call customer service, because you could be on hold a long time.

Planetary Lightspots
You'll be on the go, here, there, and everywhere, as Mars advances in Scorpio through the 27th. Be sure, though, to ease up on the gas and to emphasize patience with coworkers. If you have time, consider a long weekend getaway to a nearby destination, or stock up on books at the library and get cozy in your favorite chair.

Relationships
Mercury is your best asset in relationships this month, in both early and late October. Be alert wherever you are on the 1st and 2nd, when a chance encounter could change your life. The last ten days of October are equally promising as Mercury transits Scorpio. Ideas will flow, your mind will be extra quick, and a flash of insight could reveal clear answers to lingering questions. This time frame also has potential to bring an exciting opportunity.

Money and Success
October's New Moon on the 7th and Full Moon on the 22nd across your solar Second/Eighth House axis favor positive financial developments. Although you should continue to be thrifty, you or your mate (or both of you) could see income rise. But there's still a chance for unexpected expenses, especially while Venus is retrograde. Splurge little and save more.

Rewarding Days
1, 5, 7, 8, 10, 18, 19, 24, 29

Challenging Days
2, 3, 4, 6, 13, 20, 23, 27

Virgo/November

Planetary Hotspots

You have another easygoing month to look forward to, at least for the most part. Family communication could stumble at times with Mars in Sagittarius, your solar Fourth House. But the red planet is also great motivation to get your place in shape—clean out closets, do minor repairs, and spruce up a room or two with new and inexpensive décor. You'll also find it easy to learn the necessary do-it-yourself skills now.

Planetary Lightspots

Consider a change of scenery when the November 21 Full Moon in Taurus, your solar Ninth House, nudges your sense of adventure. Or opt for an alternative to take advantage of this influence in your travel and knowledge sector. Take a quick class, or explore online learning options you could begin in January.

Relationships

You're in the communication loop the week of the November 6 New Moon in Scorpio, your solar Third House, which Mercury transits through the 7th. Meet and get acquainted with neighbors and join coworkers for lunch and after-work gatherings. Someone close to you could offer great advice during the first two weeks of November that helps launch new endeavors for 2011. That time frame is also great if you want to host a holiday get-together.

Money and Success

Mark November 18 on your calendar. That's when Venus resumes direct motion in Libra, your solar Second House of money. Try to hold off until at least month's end to make a major purchase. But if you need domestic items or career clothing, shop while Venus is retrograde. Just be sure to confirm the store's return policy.

Rewarding Days

1, 5, 6, 10, 14, 16, 19, 20, 25, 28, 30

Challenging Days

3, 7, 9, 15, 17, 23, 24, 27, 29

Virgo/December

Planetary Hotspots
Mercury travels retrograde for the fourth time this year, switching direction in Capricorn on the 10th, and retreating into Sagittarius on the 18th. Confirm times and places so you don't miss out on a social event, and choose your words with care, especially with family members. Mercury also can trigger mix-ups with holiday gifts, and you should hold off until January if you want to purchase electronics or appliances.

Planetary Lightspots
You're drawn to the domestic scene under the December 5 New Moon in Sagittarius, your solar Fourth House, and you'll especially cherish family during this period. Make a splash for the holidays and get everyone involved in decorating and baking goodies for friends. You could receive a spectacular and surprising gift midmonth.

Relationships
Despite Mercury's retrograde status, you'll have plenty of opportunities for social events and dates with friends as Mars transits Capricorn, your solar Fifth House, from the 7th on. Do take precautions, though, to protect valuables when you're out and about, and also go with a designated driver. With Venus in Scorpio, your solar Third House, all month you'll have a way with words and can charm everyone you meet.

Money and Success
Play time is in the December forecast, but so is your career when the Gemini Full Moon/lunar eclipse on the 21st lights up your solar Tenth House. Unfortunately, it won't be stress-free, primarily because of retrograde Mercury, so slow down and double-check output. But you also could make a terrific career contact that opens up possibilities for 2011.

Rewarding Days
2, 3, 8, 12, 13, 17, 20, 26, 29, 30

Challenging Days
6, 7, 14, 15, 18, 20, 21, 27, 28

Virgo Action Table

These dates reflect the best—but not the only—times for success and ease in these activities, according to your Sun sign.

	JAN	FEB	MAR	APR	MAY	JUN	JUL	AUG	SEP	OCT	NOV	DEC
Move		7, 8						17, 18			22-30	1-6
Start a class	8, 9		4, 5							21-31	1-6	2-4
Join a club		24, 25				25-30	1-8					
Ask for a raise								13		7		
Look for work	31	1-13								15-17		
Get pro advice	18, 19	17	2-5									
Get a loan	21, 22		16, 25, 26									
See a doctor			25, 26			15, 16	15, 16				26	
Start a diet			3			30	30					
End relationship				10, 11		4, 5	4, 5					
Buy clothes	16-19	9							16, 17			7, 8
Get a makeover		1						11	7, 8			
New romance	15								16, 17			6
Vacation				5-16	12-19							

LIBRA

The Balance
September 22 to October 22

♎

Element: Air
Quality: Cardinal
Polarity: Yang/masculine
Planetary Ruler: Venus
Meditation: I balance conflicting desires
Gemstone: Opal
Power Stones: Tourmaline, kunzite, blue lace agate
Key Phrase: I balance
Glyph: Scales of justice, setting sun

Anatomy: Kidneys, lower back, appendix
Color: Blue, pink
Animal: Brightly plumed birds
Myths/Legends: Venus, Cinderella, Hera
House: Seventh
Opposite Sign: Aries
Flower: Rose
Key Word: Harmony

Your Strengths and Challenges

Charming and gracious, you're a people person who puts others at ease, even in difficult situations. This makes you a natural mediator and armchair psychologist who excels at bringing others to consensus. That's partly to fulfill your own need to keep things in harmony because you function best in an environment of peace and tranquility. But peace-loving Libra can also unleash a verbal blast when pushed too far.

Indecisiveness is a Libra trait. And you definitely are indecisive at times. But what others fail to understand is the reason behind it: you see both sides of any question and thus see the merits of diverse opinions. Nevertheless, it's wise to remember that no decision is a decision and that taking a stand is often the best choice.

You're also an idea person with a sharp mind, and you view things objectively—Libras excel at planning and strategy. Many are excellent debaters. Venus-ruled Libra also has an eye for beauty and design, and rarely will you be seen at less than your best. You're image-conscious and know the value of first impressions.

Born under the sign of partnership, you dislike doing much of anything without another person at your side. But time alone can be positive in that it gives you an opportunity for reflection. And out of that comes self-knowledge, which can benefit you throughout your life.

Your Relationships

Everyone knows you're the ultimate people person, with a wide circle of friends, business contacts, and acquaintances. As much as you value these people, not one of them comes close to your deep feelings about your partner. That's only natural because yours is the universal sign of partnership. You rarely go anywhere or do anything alone, and you feel lonely whenever you're without a companion.

But it's easy for you to confuse love with infatuation because of your strong need for a soul mate. Slow down and take time to get to know a prospective partner before you dash into commitment. Then you get the best for a lifetime. Love with an Aries, your opposite sign, could be the ultimate in passion. You also might be compatible with a Leo or Sagittarius, or one of the other air signs, Gemini and Aquarius, or a fellow Libra. However, only a very special Cancer or Capricorn could capture your heart.

You have a deep sense of responsibility for your family, and although your childhood years may not have been carefree, as an adult you can appreciate their value as a learning experience. If you're a parent, you have the same sense of responsibility for your children, but also feel somewhat detached from them—at least in their younger years. When they grow into adulthood, you become fast friends.

Many of your friends are larger than life and also well-connected. You enjoy entertaining them in style, and have a talent for making each and every person you encounter feel as though he or she is the center of your universe.

Your Career and Money

You can do exceptionally well in a career that includes working with the public, and a congenial workplace environment is a must—your productivity dips without it. The freedom to be creative within an established structure works best for you on a daily basis. Career-related changes are more the norm than the exception. But you can easily stay on top of things and in sync with emerging trends by using your sixth sense.

You have wealth potential, which can be realized through wise financial management and excellent earning potential. Nearly always thrifty, you judge each purchase according to its value. If it's worth the price, you buy; if not, you move on. Overall, and more than many signs, your goal is financial security, and you can easily use your talent for money management to achieve the comfortable lifestyle you desire.

Your Lighter Side

Your intuition is a strength, and you have a talent for subconsciously absorbing miscellaneous information, which then gels into a whole. Develop and strengthen your sixth sense and quash the tendency to overanalyze the messages of your inner voice. Skip the pros and cons and go with your instincts.

Affirmation for the Year
I find strength within.

The Year Ahead for Libra

You can expect a fast-paced year at work, thanks to Jupiter in Pisces, your solar Sixth House. Even better, you and your talents will be appreciated and in demand. Just be careful about taking on too much, which is easy to do when Jupiter is involved. Keep promises and deliver quality work on time. Then your job will be a joy and you'll get all the accolades you deserve and that are now easily earned.

The Sixth House is also your wellness sector, so tap into all that enthusiastic Jupiter optimism. Get in shape, get in the habit of nutritious meals, and fill your leisure-time hours with relaxing activities. Pets may also have a prominent role in your life this year, and you could find the perfect four-legged companion when you least expect it.

Could anything be better for Libra than lucky Jupiter in Aries, your solar Seventh House of relationships? Not much! That is exactly what is in store for you June 5–September 7, just in time for a summer romance. Consider it a bonus and a preview of 2011, when Jupiter will return to Aries. Make it a three-month celebration if you're part of a loving couple. Many other people will be a part of your life while Jupiter is in Aries, and don't be surprised if you're suddenly the most popular person around.

Saturn will be in your sign much of the year, the only exception being a brief retreat into Virgo, your solar Twelfth House, April 7–July 20. Take a breather then, as much as you can, and use a few quiet evenings to look within. Reflect on what occurred late last year and earlier this year when Saturn was in your sign. What did you learn? What do you want to do with your life? What new personal directions are you ready to embrace? Even partial answers to these questions can help you choose the right path when Saturn returns to your sign in late July.

Schedule a routine checkup, but keep your options open about whom to consult. While your usual health-care practitioner might be the best choice, it's also possible you could find someone new who's even better for you. If you have a chronic condition, an alternative medical therapy might be the answer. Be careful here, though, and do this only if it's the optimum solution.

There's one very important point to remember as Saturn progresses through your Sun sign: this merger nearly always brings exactly what you deserve. That of course can be positive or negative, depending upon the actions you've taken and the decisions you've made. Also keep in mind that what you do as Saturn transits your sign will come to fruition in fourteen years, when Saturn enters Aries and your solar Seventh House. Some of the alliances you form now will be invaluable in the future.

Another key factor associated with Saturn in your sign is a focus on the past, including regrets about what might have been or things you wish you could change. That's not realistic, obviously, but the feeling is nevertheless valid. Do what's necessary to resolve issues, either within yourself or with other people, so you can move on with a fresh agenda that focuses on your personal goals and hoped-for achievements. You'll have all the determination you need to accomplish that, thanks to steady Saturn.

Like Jupiter and Saturn, Uranus spends time in two signs this year: Pisces and Aries. You've probably already experienced the effects of Uranus, planet of change, in your workplace. Even if it didn't directly impact your job, it probably affected some of the people around you. Downsizing unfortunately continues to be a possibility, especially with Saturn in Virgo again forming an exact alignment with Uranus in Pisces. These changes could occur this spring. This alignment could also be just the motivation you need to get involved in your community or a charitable organization where you can be a positive agent for change.

There may be physical changes in your work space, as well as new technology to master. Depending upon your job, you might be in a position to telecommute, at least on some days. Take precautions if you're offered this opportunity. Set up regular communication channels with coworkers and supervisors so you're not left out of the loop. You might also decide it's time to move on, to seek a more satisfying and rewarding position. That might or might not be a wise choice. Only you can decide, based on your career field, your current company, and available jobs. Restrain yourself if you're tempted to resign before accepting another position.

Expect the unexpected in relationships this summer. Uranus shifts briefly into Aries, your solar Seventh House, transiting that sign May 27–August 12, before returning to Pisces. A soul mate—

friend, romantic interest, or mentor—could suddenly appear during this time and just as suddenly disappear. This means you shouldn't take any new relationship too seriously, and it's even more inadvisable to become legally or financially tied to anyone. With these cautions in mind, look to this person as someone who has arrived for a reason, possibly to open your eyes to what could be.

The year's second Saturn/Uranus alignment occurs in Libra-Aries; thus, at least one close relationship will be disrupted or change in some way. Endings are unlikely, however, because Uranus will slip back into Pisces. Nevertheless, you'll take first steps in that direction, weighing the past against the future, your desires and needs against someone else's. It's also possible someone from the past will reappear in your life. Ask yourself if that's the best choice for you; moving on to new territory might be a healthier option.

This is only the beginning. Many relationships will change in the next seven years as Uranus transits Aries through early 2019. You'll meet exciting, stimulating people, say goodbye to others, and revamp some relationships to better suit you both.

If you haven't yet discovered the full depth of your creative energy, take that bold step this year. Neptune in Aquarius, your solar Fifth House, is all the incentive you need to explore this side of yourself. Try decorating, computer graphics, or writing. Neptune is also the planet of inspiration, so just about any hobby or sports activity will open your eyes to new vistas. And, you can inspire the same in others.

Investments, however, require caution. With Neptune in the Fifth House, it's easy to step into the illusion of quick gains. That's unlikely. Play it safe.

Pluto in Capricorn, your solar Fourth House, will trigger significant changes on the home front at some point in the coming years. (Pluto is in Capricorn through 2024.) They could be positive, such as an extensive remodeling project. Or relocation might be necessary to take charge of a relative's affairs. You also should check your home periodically for any sign of termites.

On another, deeper level, Pluto in the Fourth House challenges you to examine your childhood and family life. Deal with unresolved issues and explore the impact these experiences had (and possibly continue to have) on your life. You could also use Pluto's energy to research and discover your family's history, genealogy, and genetics. With Pluto's powerful focus and determination on your side, you could trace your roots back many generations.

What This Year's Eclipses Mean for You

Home and family are decidedly upbeat under the January 15 solar eclipse in Capricorn, your solar Fourth House. The eclipse's close alignment with Venus, also in Capricorn, makes the ensuing six-month period a positive one for home decorating, improvements, and entertaining. (An eclipse's effects last approximately six months.) Uranus in Pisces, your solar Sixth House, is also linked to this eclipse. That's positive for telecommuting or a sideline home-based business. You also might want to host a few get-togethers for coworkers and colleagues to further develop networking contacts.

The picture changes somewhat, however, with the June 26 lunar eclipse in Capricorn. This one is closely aligned with Pluto in Capricorn and Mercury in Cancer, your solar Tenth House of career and status. Power and control issues can emerge at home and on the job, and tension results from trying to balance these two areas of life. Find an outlet for stress that you can rely on when difficult days push you to the max. This eclipse also draws in the energy of Jupiter and Uranus in Aries, your solar Seventh House of relationships. If you're house-hunting, it will be tempting to spend more than you should. Don't go there, despite a belief that income will rise to meet expenses. It may or may not. This eclipse and your partner's influence could trigger a desire to quit your day job in favor of going out on your own. What seems doable at this point, however, is likely false optimism. Rather than totally nix the idea, pursue it on your own time.

You could be offered a terrific 9 to 5 career opportunity when the energy shifts to Cancer, Capricorn's opposite sign and your solar Tenth House, at the July 11 solar eclipse. With the spotlight on your career sector you could be in line for a step up; a promotion or new position is possible. Advancement may come through a mentor or someone who sees your know-how in action. Increase the odds by having a private word with a decision-maker you can trust.

December 21 brings a lunar eclipse in Gemini, your solar Ninth House of long-distance trips and education. Both business and vacation travel are possible, and any out-of-town trip could link you with an amazing person who could further your career aims.

Be prepared to be patient, though. It could be a while before this contact resurfaces. Neighbors and relatives are other potential networking sources, but don't depend on in-laws to come through for you. This eclipse also points to education as a path to career success. A single class or seminar could make all the difference and get you up to speed on the latest techniques or technology.

Saturn

If you were born between October 21 and 23, Saturn will contact your Sun from Virgo, your solar Twelfth House, from April 7 to July 20. This transit can be a challenge for Libra because it takes you somewhat out of the social loop. But that's exactly what it's designed to do. Periodically everyone gains from an opportunity to slow the pace a little, to re-center body and soul.

Placed in your solar Twelfth House of self-renewal, Saturn encourages you to look within, to reflect and review. It's about getting to know yourself, what motivates you, and what holds you back. This process began last fall when Saturn first contacted your Sun from Virgo. The conclusions you draw this spring and summer are the final steps toward launching new personal directions when Saturn in Libra contacts your Sun. You'll know you've succeeded when you enjoy your own company more than being with others—at least part of the time.

A health matter could require your attention now, or you might need to help an ailing relative. Do yourself a favor: get a checkup and fuel your body with solid sleep and a nutritious diet.

Saturn in Libra will contact your Sun **if you were born between September 22 and October 9**. This transit will be more prominent **if your birthday is prior to September 29,** because Saturn will connect with your Sun January 1–April 6 and again from July 20 through early August. Other Libras will experience Saturn's rays for only a week or two at some time between mid-August and year's end.

Although Saturn-Sun mergers have a reputation for being difficult—and they can be—there's far more involved than you may be aware of. Above all, this is a learning experience, and the more responsibility you take for yourself and your actions the more successful the outcome. And although it's not the best time for new endeavors, it is an excellent time to begin to establish new personal directions, goals, and attitudes that will set your course for at least the next fourteen years. Be prepared, though, to feel somewhat held back because achievements probably won't come quite as easily now. Persist! Saturn rewards long-term efforts.

Saturn has a reputation for diminished vitality so you may tire more easily now. This makes sleep and relaxation important, and both will help to prevent colds and flu. You'll also be working harder now (or should be), and may experience "instant karma" if you try to take shortcuts. Anything less than a solid effort is likely to result in repercussions. This too is part of Saturn's role as a teacher. In turn, you can use your knowledge and experience to benefit others by sharing your wisdom.

Uranus

If you were born between October 16 and 22, Uranus will contact your Sun from Pisces, your solar Sixth House. Your job situation will feel unsettled, part of which is the result of your need for change. Be careful what you wish for. The April Saturn/Uranus alignment could trigger exactly that, whether as a result of downsizing or because someone behind the scenes is acting against your interests. Listen, observe, and be aware of what's going on around you.

You also may decide enough is enough and that it's time to move on. Try not to act prematurely. Your best odds for securing a new position are this fall, or possibly not until January 2011. Although it might be tough to stick it out, think carefully and weigh the pros and cons. The upside of this transit is that you could be asked to take on more responsibility. If you think that's a possibility or something you want, cultivate the idea this summer by talking with all the right people.

Your Sixth/Twelfth House axis also rules wellness, so take the time to honestly assess your lifestyle and seek input from a medical professional. Lifestyle changes made now will benefit you for many years, and you'll find it fairly easy to adapt to a new routine of diet and exercise that also will help maintain your energy level when Saturn in Libra contacts your Sun.

If your birthday is September 22 or 23, Uranus in Aries will contact your Sun between May 27 and August 12. With this planet transiting your solar Seventh House you can expect unusual people and unusual interactions to occur. Some will surprise you and others will enlighten you. There will also be a significant change in at least one close relationship, or at least the suggestion that major changes are on the horizon. Rather than resisting, listen to what the other person is saying. You will learn from it because Saturn in Libra will contact your Sun at the same time and of course form an exact alignment with Uranus in Aries. So while it might be more comfortable to maintain the status quo, you should open your mind to alternative viewpoints.

On a personal level, you'll be drawn to the idea of change but fear it as well. Nevertheless, by observing and relating to other people you can adapt and eventually even embrace the new directions that Saturn and Uranus are encouraging you to pursue.

But one area should be off limits: romantic commitment. Although the thrill of new love can be irresistible, this is not the year to lock yourself in or to make major decisions concerning a relationship.

Neptune

Neptune will contact your Sun from Aquarius, your solar Fifth House, **if you were born between October 17 and 22**. This easy connection from a fellow air sign will boost your sixth sense if you welcome the energy. Take small steps at first and learn to trust your inner voice, which will speak loudly in time, especially about you and your work life.

Romantic moments will be more than memorable this year, and you'll also have a mysterious aura that people will find irresistible. Use it to further your aims, and enjoy your love life. But keep in mind that this influence will pass and someone who intrigues you now may not by the time next year arrives.

You, among all Libras, should make it a priority to find an outlet for your creativity. Express yourself in a way that's unique to you and your interests, both in your leisure time and on the job.

Pluto

If you were born between September 25 and 28, Pluto will contact your Sun from Capricorn, your solar Fourth House of home and family. Be sure your property is fully covered by insurance, especially if you live in an area prone to severe weather. Also regularly check your home for pests and other damage. If you plan to purchase property, make the offer subject to inspection, and carefully check credentials before hiring a contractor for home improvements.

This is also the year you'll have the motivation to clean out closets, basement, garage, and attic, so go to it with gusto. And don't be surprised if a new appliance sparks the incentive to redo the kitchen. It's also possible your living situation will change because a relative or roommate moves in or out. If an adult child wants to return home, set ground rules in advance and be prepared to enforce them. You might also have issues with your parents or feel the need to resolve events from the past. While that can be healthy, it's also wise to think carefully before you speak your mind.

Jupiter-Saturn-Uranus-Pluto

This unusual four-planet alignment will contact your Sun this summer **if you were born between September 22 and 26**. It involves your solar Seventh House (Jupiter and Uranus in Aries), First House (Saturn in Libra), and Fourth House (Pluto in Capricorn). Uranus, and to some extent Jupiter, are the wild cards in this lineup, and why almost anything is possible. You can be certain, however, that events will revolve around a relationship and family. And even though it may seem like outer influences are in control, this alignment is more about you and what you need to do to take charge of your life.

You could find yourself in the middle of a difficult situation involving family members, or what you considered to be a solid relationship could undergo sudden changes, possibly because of someone new in your home. Some Libras will relocate. Open, honest communication can help resolve differences, but keep in mind that there will also be power and control issues involved, as well as subtle or overt manipulation. Any attempt to force things is likely to have an undesirable outcome.

If your home life is positive and upbeat, think twice before launching a major home improvement project, which could end up a disaster either because the work is subpar or never finished. Major do-it-yourself projects can have a similar result and spark tension among family members. Most of all, cover every possible scenario with property insurance even if you believe you won't need it.

The positive side of this extraordinary planetary alignment is that you can come to terms with the past and move forward with renewed optimism. The process will not only transform you but empower you to become your own person, confident and self-assured.

Libra/January

Planetary Hotspots
Mix-ups, misunderstandings, home repairs, and family conflict are all possible this month as Saturn in Libra squares off with Pluto in Capricorn, your solar Fourth House. However, other more favorable planetary alignments and the New Moon (solar eclipse) in Capricorn on the 15th suggest the situation will be manageable. But you'll need to take charge. Play close attention to the events that occur this month; they will continue to unfold as the year progresses.

Planetary Lightspots
Take advantage of Jupiter's remaining days in Aquarius (through the 16th), your solar Fifth House, to see friends and socialize. Thanks to Venus, which arrives in Aquarius on the 18th, the trend will continue all month, creating opportunities for romance and your favorite leisure-time activities. A former love interest or an old acquaintance may reenter your life later this month. Be cautious. Despite a strong attraction you could ultimately regret a decision to reunite.

Relationships
The January 30 Full Moon in Leo, your solar Eleventh House, is a plus for friendship, but it will be partially offset by retrograde Mars in the same sign. Some relationships will be affected; others won't. Do your best to avoid potentially difficult people and situations.

Money and Success
Jupiter enters Pisces, your solar Sixth House, on the 17th, bringing with it all the promise of an upbeat year on the job. This lucky influence accents potential gains and enhances your popularity with coworkers and supervisors. But all this enthusiasm can encourage you to take on more than you should. Commit only to what you can reasonably accomplish and you'll realize Jupiter's good fortune.

Rewarding Days
1, 2, 3, 10, 11, 15, 17, 22, 25

Challenging Days
4, 5, 6, 7, 13, 20, 23, 27

Libra/February

Planetary Hotspots
February promises to be a busy yet relatively easygoing month. The exception might be a friendship, dating relationship, or group endeavor. With Mars still retrograde in Leo, your solar Eleventh House, and contacting several planets, tension will rise midmonth. Be true to yourself even if it means you walk away; but also listen to other viewpoints if only for what you can learn.

Planetary Lightspots
February ends with the Full Moon in Virgo, your solar Twelfth House, on the 28th. Kick back, indulge yourself, make relaxation a high priority. Meditation is especially beneficial now and can open the path to your inner voice. Take note of dreams and hunches.

Relationships
Despite the potential Martian conflict this month, the February 13 New Moon in Aquarius, your solar Fifth House, favors your social life. The lunar energy also nudges your creativity and encourages you to express your individuality through a hobby or another leisure-time interest. Your children may need more time and attention now, and you should get to know their friends, especially if you have teens.

Money and Success
Your work life gets another boost this month when Venus enters Pisces, your solar Sixth House, on the 11th. Enjoy the attention and your enviable position, and cross your fingers for good news at month's end. Midmonth could bring a raise or additional company benefit. But be cautious if you're romantically attracted to someone at work. Don't put your job at risk.

Rewarding Days
4, 6, 7, 8, 13, 16, 18, 26, 27

Challenging Days
2, 3, 5, 9, 12, 17, 19, 20, 23

Libra/March

Planetary Hotspots
Strained family relationships are more the norm than the exception this month as several planets contact Pluto in Capricorn, your solar Fourth House. Keeping your cool is the main challenge, and you'll have more success resolving issues at month's end when Venus and Mercury will be favorably aligned. Until then, egos and personalities will clash. It's also possible an elderly family member may require considerable help.

Planetary Lightspots
The March 29 Full Moon in Libra lights up your solar First House, giving you extra confidence to overcome challenges and strive for personal success. Do something for yourself, something that not only pleases you but that makes a statement about you, the individual.

Relationships
This month's positive relationship news is Mars, which finally resumes direct motion in Leo, your solar Eleventh House, on the 10th. Your social life will gradually pick up, and you can smooth over recent friendship difficulties—if that's your choice. Seek networking opportunities and make an effort the next few months to add new faces to your circle of acquaintances.

Money and Success
Your work life continues to be a bright spot, thanks to the March 15 New Moon in Pisces, your solar Sixth House. Beneficial planetary alignments could trigger exciting news, possibly a step up, possibly a raise. Take note of what happens this month, as these events will come full circle this fall and carry over into 2011.

Rewarding Days
4, 7, 9, 12, 13, 14, 15, 17, 26, 30

Challenging Days
1, 2, 8, 11, 16, 18, 20, 23, 25, 29

Libra/April

Planetary Hotspots
Saturn slips back into Virgo and aligns with Uranus in Pisces. A switch to a healthier lifestyle would be in your best interests with this emphasis on your solar Sixth/Twelfth House axis, especially because job stress is on the rise. Make a point to listen to the grapevine; you could pick up advance information about possible workplace changes. If you have a pet, take your four-legged friend in for a checkup.

Planetary Lightspots
Mars picks up speed in Leo, your solar Eleventh House, highlighting friendship and socializing. Stir up some interest in a day trip with pals, nights out, and cozy afternoons with your best friend. Also continue to introduce yourself to people to make the most of this marvelous networking influence.

Relationships
This is one of your best relationship months of the year, thanks to the April 14 New Moon in Aries, your solar Seventh House. Life is divine for couples, and some single Libras take love to the next level, a step closer to lifetime commitment. You'll also be on the same wavelength with just about everyone in your business and personal lives. But be cautious if you consult a professional midmonth; ask for and expect straight answers.

Money and Success
Mercury turns retrograde in Taurus, your solar Eighth House, on the 20th, which can trigger mix-ups in money matters. Double-check anything financial and postpone major decisions until later in May. At month's end, the Full Moon in Scorpio, your solar Second House, could bring good financial news.

Rewarding Days
2, 3, 8, 9, 10, 13, 14, 17, 26, 27, 30

Challenging Days
4, 5, 7, 12, 19, 22, 23, 28

Libra/May

Planetary Hotspots
Jupiter in Pisces, your solar Sixth House, accents an upbeat, optimistic environment. But that can mask the subtle signals of this planet's alignment with Saturn in Virgo, your solar Twelfth House. Be alert for any hint of restructuring or any attempt by a coworker to undermine your position. This lineup also could trigger a heavy workload and a boss who keeps the pressure on, prompting you to consider other options.

Planetary Lightspots
Fortunately, the Universe provides a great stress reliever this month: the May 27 Full Moon in Sagittarius, your solar Third House. Plan a holiday weekend trip, or enjoy bonus days at home with a best seller; invite a few neighbors to join you and yours for a barbecue. Most of all, take some time for yourself.

Relationships
Uranus switches signs, entering Aries, your solar Seventh House, on the 27th. This puts relationship changes on the summer horizon, as well as contact with many fascinating people. Venus in Cancer, your solar Tenth House, from the 19th on, boosts your popularity and also connects you with people. Tread lightly around the 23rd, however, when controlling people will be at their most difficult.

Money and Success
The May 13 New Moon in Taurus, your solar Eighth House, could fatten your bank account with a raise, bonus, or small windfall. But caution is needed with investments, which could take a downturn; don't put funds at risk, no matter what a friend tells you. If a big-ticket purchase is in your plans, comparison shop in early May but wait to buy until after Mercury turns direct on the 11th.

Rewarding Days
5, 6, 10, 11, 12, 13, 19, 24, 28

Challenging Days
2, 4, 9, 14, 18, 22, 23, 25, 29

Libra/June

Planetary Hotspots
The Capricorn Full Moon on the 26th activates your solar Fourth House of home and family, as well as your solar Tenth (career) and Seventh (relationships) Houses. All of these areas are thus prone to stress, with any one of them influencing the other two. For example, a career change could affect domestic life and close relationships, or home-related issues could trigger relationship and career challenges. Downsizing or relocation may be the cause, as could a family member. Consider other options if a relative or romantic interest asks to move in with you. Be sure your home and property are well insured.

Planetary Lightspots
Knowledge is your best asset under the June 12 New Moon in Gemini, your solar Ninth House. Get the know-how you need for a home improvement project or to boost career skills. Your mind will be extra sharp now and geared for self-learning.

Relationships
Jupiter and Uranus join forces in Aries, your solar Seventh House, this month. That increases the odds for lucky opportunities through other people. A midmonth chance encounter could change your life, so see friends around that time. One of them could be your link to good fortune. Put networking high on your priority list.

Money and Success
Take advantage of Mercury in Taurus, your solar Eighth House, through the 9th to organize financial records and update budgets. Also analyze investment and retirement accounts, and read the latest information on health insurance benefits. This is not the time, however, to make major financial decisions or changes. Gather data you can use later in the year.

Rewarding Days
1, 7, 11, 15, 16, 20, 27, 28, 29, 30

Challenging Days
3, 6, 10, 12, 17, 19, 21, 24, 25, 26

Libra/July

Planetary Hotspots
Some close relationships will be rocky as Saturn returns to your sign on the 21st and aligns with Uranus in Aries, your solar Seventh House, on the 26th. Under this influence the desire for change and independence clashes with the need for stability and security. Blending the two will be a challenge and one that's best faced head-on. Strive for compromise even though it will be tough to find the middle ground. You also should do everything possible to learn from the experience even if you ultimately decide to cut ties.

Planetary Lightspots
Venus eases into Virgo, your solar Twelfth House, on the 10th, bringing with it the desire to step out of your hectic lifestyle. Indulge yourself! Take the time you need to rest and relax, and also treat yourself to a massage or the full treatment at a day spa.

Relationships
Friends warm your heart this month with Venus in Leo, your solar Eleventh House, through the 9th. Set aside time for social events and long talks with your best pals, several of whom will offer great advice. You could reconnect with someone from the past the third week of July, and make a life-changing contact at month's end. But a dating relationship might be on shaky ground in early July. If you're a parent, your children might stretch the truth the week of the July 25 Full Moon in Aquarius. Persist until you get the true story.

Money and Success
The July 11 New Moon (solar eclipse) in Cancer, your solar Tenth House, spotlights your career sector. Do all you can to reinforce your position. Network and offer to take on extra responsibilities. If you're job-hunting, send out resumes on the New Moon.

Rewarding Days
4, 5, 7, 8, 12, 13, 17, 26, 27, 31

Challenging Days
3, 9, 10, 15, 16, 20, 21, 22, 23

Libra/August

Planetary Hotspots
July's stresses and strains are still active, along with the accompanying relationship tension. Although you'll take steps toward resolving the situation, progress will be slow and success limited as events continue to unfold. The first and third weeks of August are likely to be the most difficult, primarily because much of what occurs will be out of your control. Or, it might just seem that way. Take charge and deal with events head-on.

Planetary Lightspots
Confidence gets a big boost from Venus, which enters your sign on the 6th. Use it for all it's worth to charm people and persuade them to see things from your perspective. You also can benefit from Mercury, which turns retrograde in Virgo on the 20th. This influence will enhance your sixth sense, giving you that advantage as well when dealing with difficult people.

Relationships
Friends continue to be a bright spot in your life, thanks to the August 9 New Moon in Leo, your solar Eleventh House. Plan a few get-togethers around that time, and consider hosting a casual backyard gathering at your place. The same time frame is a plus if you want to get together with hometown friends or attend a reunion.

Money and Success
Two planetary influences put your job in high focus this month: Uranus's return to Pisces, your solar Sixth House, on the 13th, and the Full Moon in the same sign on the 24th. You can expect some changes at work, along with a fast pace, which could pay off with a nice thank-you at month's end. Take time, however, to check and re-check work, because of Mercury's retrograde status.

Rewarding Days
1, 5, 9, 14, 22, 23, 30, 31

Challenging Days
4, 6, 7, 10, 12, 19, 20, 27, 28

Libra/September

Planetary Hotspots

Although this summer's major planetary emphasis begins to wane, you can expect yet another intense round when the September 23 Full Moon in Aries activates Saturn in your sign and Pluto in Capricorn, your solar Fourth House. Family and other close relationships will thus be stressful as you try to balance your needs with those of others. And while you'll be pushed to put your desires on the back burner, that might not be the best choice even if it is the easiest. Think carefully and deeply before you act.

Planetary Lightspots

Venus advances into Scorpio, your solar Second House, on the 8th, bringing with it the potential for higher earnings. You also can find some great buys on household items if you shop sales around the 12th or 18th. Just be sure to set a budget before you go, and also ask about the return policy in case you change your mind next month.

Relationships

Mercury resumes direct motion in Virgo, your solar Twelfth House, on the 12th. Communication will gradually resume its usual flow in the following few weeks. More importantly, you'll be able to reach a decision about a relationship with the help of your inner voice and a friend's objective advice. This month could also bring you a new pet to love.

Money and Success

Jupiter returns to Pisces, your solar Sixth House, on the 8th, and joins forces with Uranus in the same sign on the 18th. This lucky duo has all the potential to trigger a job offer or a step up with your current company. Plan ahead. Make the most of the days surrounding the 18th to line up everything in your favor.

Rewarding Days
2, 5, 6, 10, 14, 19, 20, 24, 28

Challenging Days
1, 3, 4, 9, 15, 16, 23, 25, 30

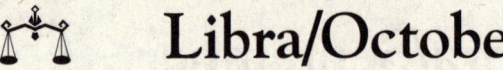

Libra/October

Planetary Hotspots

Life more or less returns to normal with only the usual daily frustrations. There is one exception, however. Venus turns retrograde in Scorpio, your solar Second House, on the 8th. Money matters will thus require attention, and you should pay bills early. You also can use this influence to advantage; organize financial records, review your budget, and boost your financial knowledge. But because Venus is your ruling planet, personal plans may be put on hold or proceed slowly.

Planetary Lightspots

Look forward to the 7th. That's the date of the New Moon in your sign, and the start of your solar year. Let this fresh energy fill you with optimistic enthusiasm; give some thought to what you want to accomplish in the next twelve months. You'll have all the determination you need to follow through on even the loftiest goals.

Relationships

October 22 brings the year's second Full Moon in Aries, your solar Seventh House. Unlike last month, this Full Moon shines brightly on your relationship sector, putting you in touch with many people. But you also can expect at least one relationship to begin to fade into the past to make way for someone new in your life.

Money and Success

Although retrograde Venus can limit income, you have three other influences going for you: Mars in Scorpio through the 27th, where it's joined by Mercury on the 20th and the Sun on the 23rd. This trio could trigger a raise or bonus, but it might be next month before you actually see the cash. Beware of impulse purchases, which will tempt you now.

Rewarding Days
7, 11, 15, 16, 17, 21, 24, 25, 26

Challenging Days
2, 5, 6, 13, 20, 23, 27, 28

Libra/November

Planetary Hotspots
November is an easygoing month with only minor exceptions. You'll want to take precautions to safeguard valuables in early November and also midmonth, when you could lose a treasured possession. Do the same with financial information. Also be cautious on the road around the 20th; better yet, catch a ride or take public transportation.

Planetary Lightspots
The 18th is a date to look forward to. That's when Venus resumes direct motion in your sign. Although it may be a few weeks before you see real results, personal endeavors will begin to regain momentum and, most of all, you'll feel like your life is getting back on track.

Relationships
You'll be at the center of the loop, a key player in the information chain from the 8th to the 29th as Mercury transits Sagittarius, your solar Third House. This is a great influence for all relationships, and especially relatives and neighbors. Give some thought to hosting a neighborhood Halloween get-together at your place, and also consider getting involved in a community project or cause. Either one will bring new people into your life, some of whom could be excellent networking contacts.

Money and Success
Your money sectors are in high focus this month under the New Moon in Scorpio on the 6th and the Full Moon in Taurus on the 21st. Both have maximum potential to boost your bank account through income, savings, and investments. However, you should postpone major financial decisions and purchases until next month, and do the same if you want to apply for credit or a loan.

Rewarding Days
4, 6, 8, 12, 13, 16, 18, 21, 22, 26

Challenging Days
3, 7, 9, 10, 17, 23, 24, 27, 29, 30

Libra/December

Planetary Hotspots

Mercury turns retrograde on the 10th in Capricorn, your solar Fourth House, and retreats into Sagittarius, your solar Third House, on the 18th. This can trigger misunderstandings with family members and other relatives, as well as mechanical problems. Check your car battery and consider a replacement if it's older. Also periodically check household appliances. Try to hold off until January if you plan to purchase a big-ticket item, and confirm the return policy if electronics are on your holiday gift list.

Planetary Lightspots

Your adventuresome spirit comes alive this month with the New Moon (December 5) in Sagittarius and the Full Moon/lunar eclipse (December 21) in Gemini. If you want to get away from it all, choose a nearby destination for a long New Moon weekend. That's a better choice than long-distance travel near the Full Moon, which will be prone to a delay or cancellation.

Relationships

Family relationships run the gamut from stressful to upbeat as retrograde Mercury disrupts communication. Mars, which enters Capricorn on the 7th, could trigger conflict off and on all month. Think peaceful thoughts and keep your cool, both of which will help prevent things from escalating. And don't hesitate to end conversations rather than to argue with difficult people who are unlikely to change their minds.

Money and Success

Establish a holiday gift budget. Otherwise, Venus in Scorpio, your solar Second House, could encourage you to go overboard. The positive of this placement is you could receive a year-end bonus and many wonderful gifts.

Rewarding Days
1, 4, 5, 8, 10, 13, 19, 23, 29

Challenging Days
7, 11, 14, 15, 20, 21, 25, 27, 28

Libra Action Table

These dates reflect the best—but not the only—times for success and ease in these activities, according to your Sun sign.

	JAN	FEB	MAR	APR	MAY	JUN	JUL	AUG	SEP	OCT	NOV	DEC
Move	16–18			5, 6			23–25		16, 17			
Start a class			25, 26	2–4			21, 22			11, 12	7–30	1–6
Join a club			25, 26			15, 16	23–27				26, 27	2, 3
Ask for a raise			4, 5									
Look for work	18, 19	14–28	2–16									
Get pro advice		17, 18	18–31	1								
Get a loan		19		15, 16		7		2, 3				
See a doctor		1, 15	15					11	7			
Start a diet				10, 11	7–9	4, 5	28, 29					
End relationship					10, 11	6, 7	3–5, 31	1	23, 24			
Buy clothes	20–31	1–10			5, 6							
Get a makeover		2, 3				19, 20		13, 14				
New romance		13, 14				28, 29						
Vacation						10–20						

SCORPIO

The Scorpion
October 22 to November 22
♏

Element: Water

Quality: Fixed

Polarity: Yin/feminine

Planetary Ruler: Pluto (Mars)

Meditation: I can surrender my feelings

Gemstone: Topaz

Power Stones: Obsidian, amber, citrine, garnet, pearl

Key Phrase: I create

Glyph: Scorpion's tail

Anatomy: Reproductive system

Color: Burgundy, black

Animal: Reptiles, scorpions, birds of prey

Myths/Legends: The Phoenix, Hades and Persephone, Shiva

House: Eighth

Opposite Sign: Taurus

Flower: Chrysanthemum

Key Word: Intensity

Your Strengths and Challenges

Determination is a major Scorpio strength, and you ignore obstacles and hurdles that others see as insurmountable. When you set your mind on a goal, you rarely waver, even if it takes years to achieve. That's a real plus in life, but there are times when stubbornness replaces determination. Learn to recognize the difference and to go with the flow when it will get you further.

Pluto, your ruling planet, gives you the willpower to outlast just about anyone. But at times you don't know when to quit, especially when doing anything physical. Pace yourself and stop short of exhaustion. Even though you feel invincible, you're really not.

Scorpio's aura is mysterious and magnetic, and your unwavering gaze mesmerizes many. One look from you says it all, and you also can sum things up in a few words. You're shrewd, intuitive, and creative, a combination that gives you the edge in many situations, and you act only after you have all the facts and when it's to your advantage. This of course contributes to your reputation for secrecy, and what you hear in confidence is seldom repeated.

Pluto and Scorpio are the planet and sign associated with transformation. You use this power on a practical level, personally, and in your career. No one can match your ability to take what appears to be junk and create something useful, and you're a change agent in your daily work. Personally, you periodically reinvent yourself, whether to change your appearance or gain new skills.

Your Relationships

You're choosy about the people you invite into your life, especially friends. Most Scorpios have many acquaintances but only a few friends who enter their inner circle after proving themselves and earning trust. Some may be former or current coworkers or people you encounter through volunteer activities or a professional organization. You're also more comfortable with people you know well than at social events.

Love, romance, and passion are essential in your life, and you need the deep emotional connection that comes from a loving partnership. At times idealistic in matters of the heart, you're also possessive because you can't imagine life without the one you love. And that can result in unfounded jealousy, which can erode your relationship over time. Love your sweetie with your ultra-romantic heart and have faith in your commitment. You could make a sensational match with Taurus, your opposite sign, or one of the other water signs, Cancer and Pisces. Virgo and Capricorn are often compatible with Scorpio, but Leo or Aquarius may not be the best choice.

Your family relationships are an unusual mix of cool detachment and unbreakable bonds. Some are your best friends and you rarely have contact with others. But as a parent, you're totally a soft touch, giving in more often than not. That might make you—and your children—feel good for the moment, but ask yourself if you're really doing them a favor. Firm guidelines and teaching them life values will serve them better in later years.

Your Career and Money
Scorpios value their privacy. Yet you also enjoy center stage and strive for an influential position in the world. Your career is an excellent outlet for your leadership ability because you can inspire others and bring people together to achieve mutual aims. Many Scorpios are found in executive positions. In your daily work you thrive in a fast-paced environment that values initiative. But be cautious about pushing yourself and others too hard.

Lucky in money matters, you have the potential for high earnings and sizable assets. Budgeting, knowledge, and information are your keys to financial success, as are conservative long-term investments. At times you have a tendency to splurge, so wait a day or two before making a major purchase; you could just as easily change your mind.

Your Lighter Side
You're a people person at heart, although most you encounter don't fully realize it. That's because you share your deepest thoughts and feelings with only a select few. You're also kind-hearted and do all you can for those who touch your soul—friends, strangers, and family members.

Affirmation for the Year
I embrace fresh goals and new opportunities.

The Year Ahead for Scorpio

Welcome to a year of play! Not exactly, but you will have an active social life, thanks to Jupiter in Pisces, your solar Fifth House, much of the year. Jupiter expands whatever it touches, so 2011 also has the potential to be memorably romantic whether you're single or committed. The odds are that Scorpios on the dating scene will have many admirers to choose from.

This could be your lucky year if you're hoping for an addition to the family, and parents of children of all ages will delight in many happy moments. But you'll also have a tendency to spoil those you love, so begin the year with a budget and stick to it. You might enjoy getting involved in your children's activities, and could have success heading up a fundraising campaign for their sports team or after-school club.

Leisure-time activities will be high on your priority list, so 2010 is also your opportunity to join a sports league or to learn a new sport like tennis, racquetball, or golf. Or organize a neighborhood walking group or dive into a new hobby, something you've always wanted to try. Find a way to express your individual brand of creativity.

Jupiter visits Aries, your solar Sixth House, June 5–September 7, to give you preview of 2011, when the lucky planet returns to this sign. This can be a real plus for your work life. You may have a chance to earn some extra money, possibly related to an exciting opportunity. Just be sure not to take on more than you can handle and confirm that you will in fact be paid. Think carefully if you're offered a job with another company; it may or may not last.

Saturn begins and ends the year in Libra, making a final visit to Virgo, April 7–July 20. During its time in Virgo, your solar Eleventh House of friendship and group activities, you may reconnect with a friend from the past. But this person might be less interested in you than in a favor you can grant. You'll part ways with some friends because of a lifestyle change or relocation, and you may drift away from others.

This is a good time to be involved in a professional or charitable organization as long as you do only your fair share. You might also decide it's time to cut ties with a club or organization because it no longer interests you or you simply want more time for other activities.

Although it might be tough to walk away, push yourself if you know it's truly the right choice. Complete projects and commitments before you do. Part of your motivation to step away from friends and group activities is Saturn's emphasis on Libra, your solar Twelfth House of self-renewal. Time alone is appealing and you're ready to simply enjoy your leisure hours rather than have them filled with responsibilities and organized activities. You're also ready for at a least a partial breather from the high-level people contact of the past few years.

Saturn's main purpose, as it transits your Twelfth House, is to prepare you for its arrival in your Sun sign. Review and reflect upon what you've accomplished in the past seven years, as well as where you feel you've fallen short. You can learn from both. Put things in perspective, resolve regrets and issues, and celebrate successes. This exercise will help you learn more about yourself and your strengths and weaknesses so you're better prepared to capitalize on the qualities that will maximize success.

Like Jupiter and Saturn, Uranus also transits two signs this year: Pisces and Aries. Placed in Pisces, your solar Fifth House, much of the year, Uranus encourages you to take a fresh look at yourself and your life. Desired changes are more easily achieved now and you'll have the inspiration and determination to take charge and get in shape, learn healthy cooking, or almost anything else that will make you feel terrific about yourself.

Mix your talent for transformation with the year's abundant creative energy and express your individuality by redoing a room, the garden, or your whole house! If you have an innovative streak, it's likely to emerge this year. Who knows? You might invent a new must-have gadget. Do be cautious, however, with investments. Uranus can bring you riches but also reduce your net worth. Entrepreneurial ventures are chancy at best, but you could luck into a nice win in September when Jupiter and Uranus join forces in Pisces.

April brings an exact alignment of Saturn in Virgo with Uranus in Pisces. A friend could lean on you for a loan at that time, but repayment is unlikely. So if you go ahead, consider it a gift. And don't hesitate to say no if an organization expects a big donation or asks you to take the lead in a fundraising project; you could regret doing either one. Curb your generosity. Protect your resources.

Uranus transits Aries, your solar Sixth House, May 27–August 12. A job change—or several—is possible and even likely in the next seven years, and you may eventually telecommute. These changes could be triggered by company decisions, but they're just as likely to be initiated by you. But try not to get caught in the job-hopping trap in your search for the ideal position, which in reality doesn't exist.

Work-related changes may occur this year, although they may not directly involve you. An increased workload is possible, as is a new work location, and new technology may in some way impact your job. Saturn will again align with Uranus, but across your solar Sixth-Twelfth House axis, so your work life will feel unsettled. It's also wise to listen, observe, and use your sixth sense to tune in on what might be happening behind the scenes. Secrets and personal information are best kept confidential.

This is also your wellness axis, so do yourself a favor and schedule medical, dental, and eye checkups. And even though you're not usually high-strung, this Uranus transit could be the exception. Plan ahead. Find a stress reliever that works for you and get in a new daily habit. You'll feel and sleep better, in addition to being more productive and upbeat.

Neptune continues its long transit through Aquarius, your solar Fourth House, the sign it entered in 1995. This creative planet complements Uranus, encouraging you to use your creativity in household projects. Home will also be your ideal location to get away from it all this year, especially this summer. Even if you usually take a vacation trip, consider a week or two at home instead; it probably will be more relaxing. This year is also great for entertaining close friends, and you'll prefer that to the social scene at times.

You'll experience every facet of family relationships. Some relatives will inspire you and others will confuse or disappoint you. They're merely a microcosm of the world, and can aid your understanding of human nature. You'll be especially involved in your children's lives, whether they're youngsters, teens, or adults. Whatever their age, you can do much to encourage them, and they'll value your support more than you'll ever know.

Deep thinking accompanies Pluto's Capricorn transit. Placed in your solar Third House, this powerful planet could motivate you to study a subject, or to sit down at the computer and write a book. But Pluto here is more about looking within to discover the hidden side of yourself. Self-help books can spur your thinking and be an asset in jumping any hurdles that hold you back.

The Third House is also associated with siblings and neighbors. You'll be more involved with these people, and you could succeed in championing a community cause or project to improve your surroundings. As a catalyst for change, the effort you put into it will be rewarded many times over. You're also in a position to influence people one-on-one, to be the cheerleader who makes a difference and ultimately changes someone's life for the better.

What This Year's Eclipses Mean for You

There are four eclipses in 2010, two solar and two lunar. The effects of each are active for approximately six months. Capricorn is the sign of both a solar and lunar eclipse occurring in January and June, respectively. With the emphasis on your solar Third House, these eclipses spotlight communication, quick trips, neighbors, and relatives. This is also your learning sector.

The January 15 solar eclipse has all the potential to trigger news, and possibly a lucky win or a new romance. For that you can thank the eclipse's alignment with Venus, also in Capricorn, and Uranus in Pisces, your solar Fifth House. You'll be on the same wavelength as your children and partner, and have the opportunity for a winter or spring vacation in a sunny location. Positive thinking is an added plus, and your people skills will be at their best, encouraging others to respond favorably to your requests.

The June 26 lunar eclipse in Capricorn isn't quite so upbeat, however. It's aligned with Pluto in Capricorn, Mercury in Cancer (Capricorn's opposite sign and your solar Ninth House), and Jupiter and Uranus in Aries, your solar Sixth House. This raises the possibility of legal matters, power plays at work or with relatives, and potentially expensive mechanical problems. Be sure to ride with a designated driver when you're out socializing. But you also could be offered a terrific job or an opportunity to complete or advance your education.

July 11 brings a solar eclipse in Cancer, reinforcing the June eclipses emphasis on legal matters and education. Its alignment with Mars in Virgo, your solar Eleventh House of friends and groups, makes networking an excellent path to success. Consider how advanced training could enhance your job skills, as well as involvement in a professional organization where you can connect with people in your career field. Travel for business or pleasure is also in the forecast, and you could make a job connection while attending a conference or seminar.

The year's eclipse energy shifts away from Cancer/Capricorn on December 21, with the lunar eclipse in Gemini, your solar Eighth House of joint resources. Personal as well as family income could rise or fall as a result of this eclipse because it's aligned with Jupiter and Uranus in Pisces, your solar Fifth House. Extra expenses involving your children are possible, but you might also net a windfall. With retrograde Mercury also aligned with the eclipse, you should postpone major purchases and financial decisions at least for a few months. Money owed to you may be delayed.

Saturn
If you were born between November 19 and 22, Saturn will contact your Sun from Virgo, your solar Eleventh House, and connect you with longtime friends and other people who can stimulate your thinking. You may meet someone now who's a soul mate or with whom you have a karmic tie, and there could be a similar link with a charitable organization. Because this is Saturn's final transit in Virgo (April 7–July 20) you should resolve any differences with friends and may end your involvement with a group—after fulfilling any responsibilities. You'll have an especially meaningful experience if you devote your time and talents to this organization, and might feel compelled to help those less fortunate before moving on.

If you were born between October 22 and November 8, Saturn in Libra will contact your Sun from your solar Twelfth House. This transit will be more prominent **if your birthday is between October 22 and 27** because Saturn will make two connections—January 1–April 6 and again from July 21 until early August. This transit is about self-understanding, which begins with getting in touch with your inner voice. Ask yourself what you fear, what you wish and hope for, and what motivates you and holds you back. Be honest with yourself, even if it takes some time to reach that point. Then set plans in motion to deal with and resolve those things you identify as self-limiting. Also give meditation a try; it will help calm your mind and let subconscious thoughts emerge.

You may be in contact with someone who is hospitalized (or otherwise institutionalized). Or it might be necessary to manage the affairs of an elderly relative who needs an advocate. Doing your best for this person (or people), although stressful, will ease your soul and possibly satisfy a karmic debt. The Twelfth House is also known as the hidden side of life. So in many respects what you resolve now will free this side of you so that you can appreciate and value relationships and realize that security comes from within, not from another person.

Uranus

If you were born between November 15 and 22, Uranus will contact your Sun from Pisces, your solar Fifth House. Change, although rarely your favorite word—unless it's your idea!—will come easier this year. In fact, you might find yourself not only open to the idea but welcoming it. Open yourself to new experiences, new people, new thoughts, and a new you. Recreate yourself into a healthier body and lifestyle, and explore subjects that catch your interest. This is a great year to take up a creative hobby, for it will encourage self-expression and self-understanding.

You will benefit as much from your children. Get to know them as individuals and help them discover their many talents and interests, which may not be the same as yours. You'll also be more accepting of their need to assert their independence. But you'll want to keep a closer eye on your children and who their friends are if Saturn in Virgo also contacts your Sun (**Scorpios born November 19–22**). Find the middle ground between too much and too little structure and programming, and apply the same philosophy to your own life.

New love is in the forecast for some Scorpios this year, although it may be merely pleasant and lack passionate zing. That's okay. This person, like many others, will serve a purpose: to help you better define what you want or appreciate in a partner.

If you were born October 22 or 23, changes at work will be unsettling as Uranus contacts your Sun from Aries, your solar Sixth House (May 27–August 12). Although a major event such as downsizing could occur, you may experience this transit as an indefinable shift in energy that will have you feeling out of step. A new coworker or supervisor could disrupt the workplace environment or you might be asked to take on additional responsibilities not to your liking. These and other changes such as work hours and location could prompt you to look elsewhere, but this isn't the most opportune time for a job change. The accompanying tension could lower your immune system and make you prone to a cold or flu. Find a stress reliever that works for you and practice it daily.

Neptune

If you were born between November 16 and 22, Neptune will contact your Sun from Aquarius, your solar Fourth House of home and family, beginnings and endings. On a practical level, your home could suffer damage from a leaky pipe or appliance or severe weather. Flooding is possible, so be sure you have adequate insurance.

Communication with some family members will be difficult and confusing, almost as if you're speaking a different language—and in a sense you will be. You'll also find it tough to understand their perspective and grasp their motivations. In addition, you may become a caregiver for an elderly or disabled relative.

But this transit is even more about you and your search for identity. Family and childhood are key components here, and you can learn much about yourself by examining their impact (past and present) on your personality, needs, goals, and desires. This will help counteract the disillusionment this transit can bring. Be patient if you feel directionless. This too is part of the Neptune transit. Question yourself about this feeling and zero in on these basic issues. Then, when Neptune departs, a clear picture of your new direction will emerge. For all these reasons it's wise to postpone major life decisions rather than pushing yourself to get back on track. Take note of your dreams; they can be insightful.

Pluto

Pluto, your ruling planet, will contact your Sun **if you were born between October 26 and 28**. This favorable connection will help you take charge in almost any situation, and you'll have the willpower to transform your image and how the outer world views you and your talents. You really can change whatever you wish about yourself this year, whether that's a healthier lifestyle or a new look, or developing your communication and public-speaking skills.

This year's Pluto influence extends to relatives, especially siblings and neighbors. Make an effort to resolve any long-standing issues with these people, with whom you'll be more involved. You also could be a positive force for change in your community.

Learning has a role with Pluto in your solar Third House. Your powers of concentration will be strong and you'll be able to focus and master whatever you study. Just be cautious about letting a new subject of interest consume your every day and your life.

Jupiter-Saturn-Uranus-Pluto

If you were born between October 22 and 26, these four planets will contact your Sun this summer. This unusual lineup involves Jupiter and Uranus in Aries (Sixth House), Saturn in Libra (Twelfth House), and Pluto in Capricorn (Third House). Change is the keyword here and events will unfold quickly and unexpectedly.

Your job could be at risk because of global conditions, or a shake-up on the job could boost your work load. Be cautious if a job opportunity comes your way; all may not be as it seems on the surface. Going into business for yourself or with a family member, however, is a risky proposition. But a work-related event could be just the incentive you need to begin or complete your education or to fine-tune your marketable skills. This will pay off within a year.

Get a checkup if you haven't had one recently or if a health issue concerns you. Also be alert on the road and always go in the company of a designated driver.

Scorpio/January

Planetary Hotspots
The January 15 New Moon (solar eclipse) in Capricorn, your solar Third House, accents good news and a welcome opportunity as Mercury, also in Capricorn, resumes direct motion the same date. But it could be accompanied by someone's hidden agenda. Ask questions and use your intuition and research talents to get the facts behind the gloss. More will be revealed as the year unfolds.

Planetary Lightspots
Lucky you! Jupiter enters Pisces, your solar Fifth House, on the 17th, highlighting creativity, romance, leisure time, and children. All of these will have a greater influence in your life in the coming months. So take a little time now to explore possible outlets—hobbies, sports, your children's activities. An exciting new love interest (or maybe several!) is on the horizon for some singles.

Relationships
You're drawn to the domestic life from the 18th on when Venus is in Aquarius, your solar Fourth House. Comfy evenings and weekends with family will be both relaxing and sentimental; plus, the quieter pace will be good for you. Host a casual get-together for close friends the third or fourth weekend.

Money and Success
Time at home is the ideal stress reliever for what at times will be a frustrating career month. Projects will stall and decisions will be put on hold. Wait it out with patience and try not to push, which will only increase your frustration level and create turmoil. In the meantime, take advantage of the bonus hours to catch up on any backlog, and to plan, organize, and review.

Rewarding Days
1, 2, 4, 8, 14, 18, 19, 24, 28

Challenging Days
3, 6, 13, 16, 20, 23, 27, 30, 31

Scorpio/February

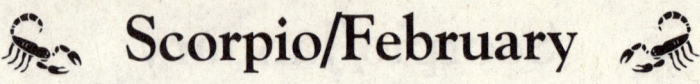

Planetary Hotspots
February promises to be a relatively easygoing month with the exception of Mars, still retrograde in Leo, your solar Tenth House. Expect delays and frustration to continue off and on, and do your best to avoid bringing the stress home with you. Also be cautious about voicing your true opinions, even with someone you believe you can trust. That could backfire one way or another in the coming months. Keep thoughts to yourself.

Planetary Lightspots
Home life continues to be a bright spot, thanks to the February 13 New Moon in Aquarius, your solar Fourth House. Communication flows and you're in sync with family members from the 10th on when Mercury is in Aquarius. The week of the New Moon is great for domestic projects, cleaning, and entertaining. Redo a room if you can take a few days off.

Relationships
Friends, social events, and fun are in the February forecast, with Venus in Pisces, your solar Fifth House, from the 11th on. That's a great influence for dating and meeting potential love interests. If you're a parent, your children will delight you and you could have much to celebrate. Plan a family outing or day trip midmonth.

Money and Success
The February 28 Full Moon in Virgo, your solar Eleventh house, accents networking and goal-setting. Get out and about with friends and connect with new people. Also make a date with a wise pal who's a good listener and seek advice on how best to achieve your worldly goals this year. Some of what you hear will surprise you, and you may realize you're farther along than you thought. Deep thinking yields more answers.

Rewarding Days
1, 4, 6, 10, 13, 14, 15, 16, 24, 25

Challenging Days
2, 3, 5, 7, 9, 12, 17, 19, 20, 23

Scorpio/March

Planetary Hotspots
Several planets in Aries, your solar Sixth House, clash with Saturn in Libra and Pluto in Capricorn this month. You can thus expect ongoing stress and strain in your daily work, as well as the potential for conflict with coworkers or supervisors. Try not to make snap decisions or act in haste, and do your best to steer clear of power plays. Also be alert for anyone who might try to play the blame game at your expense.

Planetary Lightspots
Take a well-needed break at month's end when the March 29 Full Moon highlights your solar Twelfth House of self-renewal. Even a few days to enjoy your own company and the people you love will refresh you, body and soul. You might even consider a long weekend at a nearby destination for an uplifting change of scenery.

Relationships
Fun times, fun people, and social events continue in the spotlight, thanks to the March 15 New Moon in Pisces, your solar Fifth House. Plan an outing midmonth, when the odds favor a lucky connection—romantic possibility, networking contact, or new friend. With Mercury in Pisces through the 16th, you'll have all the right words to charm just about anyone.

Money and Success
Good news! Mars in Leo, your solar Tenth House, resumes direct motion on the 10th. It will be several weeks, maybe early April, before career matters get up to speed, but even a small amount of progress will encourage you to stick with it. A meeting near the 20th could trigger insights into the future and what might develop career-wise later this year.

Rewarding Days
4, 9, 10, 14, 15, 19, 26, 27, 28

Challenging Days
2, 8, 11, 16, 18, 20, 23, 25, 29

Scorpio/April

Planetary Hotspots
Your solar Eleventh-Fifth House axis is this month's hotspot as Saturn returns to Virgo and aligns with Uranus in Pisces. A strained friendship or dating relationship could end, or you might cut ties with a group or organization you've been involved in for some time. If you're a parent, keep a close eye on your children and their activities and friends, especially if you have teens asserting their independence.

Planetary Lightspots
Look forward to the 28th. That's the date of this month's Full Moon in your sign, which is sure to enhance your magnetic aura. Use it to attract people into your orbit, trade favors, and build ties. Also give yourself a goal checkup. Evaluate, review, and revise progress toward personal goals you set in motion earlier this year or late last year.

Relationships
You're in sync with just about everyone this month as Venus transits your solar Seventh House through the 24th. With a little effort you can even smooth things over before conflict erupts, or at least walk away feeling okay about the outcome. There is, however, one potential blemish in this otherwise upbeat picture: Mercury turns retrograde in Taurus on the 17th. That can trigger misunderstandings, so choose your words with care.

Money and Success
Your work life gets a boost of energy from the April 14 New Moon in Aries, your solar Sixth House. A creative approach adds to your success, and you'll want to go all out to set yourself up for potential gains this summer.

Rewarding Days
1, 6, 10, 11, 14, 15, 16, 20, 27, 29

Challenging Days
2, 4, 7, 12, 19, 22, 23, 25, 28

Scorpio/May

Planetary Hotspots
Jupiter in Pisces aligns with Saturn in Virgo across your solar Fifth/Eleventh House axis. This planetary alignment will put you in touch with new people, but also could prompt you to finally distance yourself from a friend. All these new faces will make you smile, and among them could be someone who becomes a lucky charm this summer. However, take your time getting to know these people and don't share any secrets no matter how comfortable you feel.

Planetary Lightspots
Venus enters Cancer, your solar Ninth House, on the 19th, bringing with it the desire for fresh scenery. Opt for a weekend away close to home rather than a long-distance trip, or attend a reunion, where you could meet your soul mate. If you're part of a couple, put your heads and hearts together and plan a romantic summer vacation.

Relationships
Relationships couldn't be better this month after Mercury turns directs in Taurus, your solar Seventh House, on the 11th, followed by the New Moon in the same sign on the 13th. That's simply superb for love and romance, and commitment is in the forecast for some Scorpios. This is also a good time for professional consultations because you can get straightforward answers. Do, however, take the time to check references.

Money and Success
Finances are mostly status quo with the May 27 Full Moon in Sagittarius, your solar Second House. But extra expenses are possible midmonth, and you need to think carefully about a financial request from a friend or organization. Weigh that decision against the needs of you and yours, including your children's activities and college fund, which might be a better investment.

Rewarding Days
3, 7, 8, 12, 17, 24, 27, 28

Challenging Days
2, 5, 9, 13, 16, 18, 22, 23, 29

Scorpio/June

Planetary Hotspots
You'll want to avoid travel at month's end if at all possible, when the June 26 Full Moon (lunar eclipse) in Capricorn, your solar Third House, clashes with several planets. Delays and cancellations are possible as well as lost luggage. The planetary energy also urges caution on the road, and you might want to play it safe and take public transportation or catch a ride to work and to social events the last week of June.

Planetary Lightspots
Your popularity is on the rise from the 14th on when Venus enters Leo, your solar Tenth House. Enjoy your attention-getting status and use it to advantage at work and at play. Even better, you'll have the opportunity to mix and mingle with decision-makers. One of them could be your key to advancement within the next few months.

Relationships
You could reconnect with an old friend in early June, when you'll also enjoy a slow-paced day with your best pal far more than attending a meet-and-greet social event. Listen to this friend's wise advice about a confusing family situation. Let your intuition be a guide if you question a serious dating relationship at this time. It's probably on target, and this is not the time to establish a home with a romantic interest.

Money and Success
Jupiter enters Aries, your solar Sixth House, on the 6th, where it merges with Uranus two days later. This lucky influence could trigger a fabulous job offer or a step up. Curb your excitement just a little, though. All might not be quite as rosy as it seems. Get the facts and back them up with your own research.

Rewarding Days
4, 5, 8, 9, 13, 14, 15, 17, 21, 22, 27

Challenging Days
3, 6, 10, 12, 19, 23, 24, 25, 26

Scorpio/July

Planetary Hotspots

Saturn returns to Libra on the 21st, and five days later aligns with Uranus in Aries, your solar Sixth House. This difficult combination could trigger workplace changes, including downsizing. It may or may not directly affect your job. Instead, you could be asked to take on added responsibilities. Although this may not seem like an opportunity—just more hard work—it will be if you use it to advantage. Think carefully, however, if you're suddenly offered a job elsewhere; it could end as quickly as it begins. Also schedule a checkup if it's been a while since your last one.

Planetary Lightspots

The Universe offers you a great antidote for this month's job stress: the July 23 Full Moon in Aquarius, your solar Fourth House. Delight in family time, and do something creative to enhance your home. Most of all, simply enjoy lazy evenings and weekends in the space you call your own.

Relationships

Friends warm your heart this month as Venus travels in Virgo, your solar Eleventh House, from the 10th on, where it's joined by Mercury on the 27th. Make networking a top priority because a pal could be your link to good fortune. However, say less rather than more to neighbors. Keep personal business private.

Money and Success

Your status rises this month, thanks to Venus in Leo, your solar Tenth House, through the 9th, the date Mercury arrives in Leo. Connect with people and invest your energy and determination in work. The effort is well worth it and could pay off nicely next month. If you're job-hunting, send out as many résumés as possible, and you could realize success in early August.

Rewarding Days

1, 2, 6, 7, 11, 14, 15, 24, 29

Challenging Days

3, 9, 10, 16, 21, 22, 23, 30

Scorpio/August

Planetary Hotspots
Job-related tension continues this month as planets in Aries and Libra clash with Pluto in Capricorn. Be especially cautious about what you say to whom, and do your best to avoid difficult, controlling people. Someone who appeared a friend and supporter last month could reveal his or her true self by attempting to undermine your actions. You'll also have some tough decisions to make regarding your work life. Try to stay focused on the big picture rather than the daily frustration because this influence will pass in the next few months.

Planetary Lightspots
Venus eases into Libra, your solar Twelfth House, on the 6th. See this as a welcome break from the outer world and your opportunity to re-center, body and soul. If your thoughts drift to the past, releasing regrets, take action to resolve them, which you can do with great success now.

Relationships
The August 24 Full Moon in Pisces energizes your solar Fifth House of romance, children, and leisure time. Make all a part of your life in the following two weeks, including spending extra time with your kids. They'll love you for it as much as you'll delight in their company. Some Scorpios connect with a new love interest at the least expected moment.

Money and Success
Your career sector gets a burst of fresh energy under the August 9 New Moon in Leo, your solar Tenth House. Use it for all it's worth, whether you're in an established position or looking for a new one. And remember these upbeat days when frustration tries to get the best of you later in August.

Rewarding Days
2, 3, 11, 14, 15, 16, 21, 29, 30, 31

Challenging Days
4, 6, 10, 12, 13, 19, 20, 27, 28

Scorpio/September

Planetary Hotspots
Although it might be tough to deal with more events related to this summer's difficult planetary alignments, you're on the home stretch. The energy peaks at the September 23 Full Moon in Aries, your solar Sixth House, when the lunar energy activates Saturn in Libra and Pluto in Capricorn. Continue to be cautious about what you say, use your sixth sense to get the edge on the competition, and try not to get drawn into power plays. And be careful on the road.

Planetary Lightspots
Smile! Venus enters your sign on the 9th, an influence that's sure to enhance your magnetic charm. With it comes an extra level of charisma that attracts even more people into your orbit. Make the most of it in every area of your life, especially with your mate or dates and in every one-on-one conversation. You have what it takes!

Relationships
Your social life takes off under the September 18 New Moon in Virgo, your solar Eleventh House, making this month great for get-togethers with friends. You could step up to a leadership position with a group, and networking continues to be one of your best routes to success. Some singles will suddenly realize that a friendship could be much more. If you're among them, follow your heart and get things started.

Money and Success
Luck is with you, thanks to Jupiter, which returns to Pisces on the 8th and joins forces with Uranus on the 18th. Take a chance on the lottery, or better yet, add a dollar to a group purchase. You could come out a winner. Investments require caution, however. What seems like a winner could be exactly the opposite.

Rewarding Days
2, 7, 8, 11, 12, 17, 20, 22, 27

Challenging Days
3, 4, 9, 15, 16, 18, 23, 25, 26, 30

Scorpio/October

Planetary Hotspots
You'll experience minor frustrations this month with Venus turning retrograde in your sign on the 8th. Some relationships, especially romantic ones, will cool, and you may distance yourself from people for no obvious reason. Doing this has its upside, however; it will give you a chance to evaluate specific ties and your overall approach to relationships. With Mars in your sign through the 27th, you'll also be impatient at times with yourself and other people. Remind yourself to go with the flow.

Planetary Lightspots
Your sixth sense is particularly active now, with Mercury in Libra, your solar Twelfth House, from the 3rd to the 19th. Enhancing the influence is the October 7 New Moon in the same sign. Set aside time each day to meditate or sit quietly. This will help your inner voice emerge.

Relationships
Despite the influence of Venus and Mars, you'll be in sync with most people from the 20th on, when Mercury is in your sign. In fact, you could make a fortunate contact during the last ten days of the month. Just don't take everything you hear on faith. Use the best of your Scorpio traits to ask questions and confirm facts so enthusiasm doesn't get the best of you.

Money and Success
Your work life is satisfying and rewarding as the October 22 Full Moon in Aries lights up your solar Sixth House. The lunar energy is ideal for completion, so try to wrap up projects and tasks during the following two weeks. Step away periodically, though, to refresh your perspective; intense focus could cause you to miss the obvious.

Rewarding Days
1, 5, 8, 10, 18, 19, 24, 29

Challenging Days
2, 3, 4, 6, 13, 17, 20, 23, 27

Scorpio/November

Planetary Hotspots
Life perks along in November with only minor distractions and irritations. The main challenge you'll have to contend with is the feeling that you're off-pace and out of sync with the world and even your life. But you can use this planetary energy to advantage by delving into your subconscious. Images will emerge at odd moments, and the more you accept your limitations the more inspired you'll feel.

Planetary Lightspots
There's no question about it. You are this month's lightspot! For that you can thank the November 6 New Moon in your sign, which signals the start of your solar year. Tap into the fresh lunar energy and make it your own as you plan what you want to achieve in the next twelve months.

Relationships
November brings a beautiful Full Moon in Taurus, your solar Seventh House, on the 21st. You'll find it easier than ever to attract people into your orbit, and others will seek your assistance and advice. Give both freely. The lunar energy sparks a new romance for some singles, and couples can rediscover all the reasons they fell in love. However, if you've been wavering about whether to take a relationship to the next level, this Full Moon may trigger a decision.

Money and Success
Money matters are generally positive this month, with the potential for increased income. There is potential danger here, however, with Mercury and Mars transiting Sagittarius, your solar Second House. You'll be in the mood to splurge. Remember that when you shop, and establish and stick to a holiday gift budget. With that thrifty mindset, you can have great success searching for and finding bargains.

Rewarding Days
1, 2, 5, 6, 10, 14, 16, 19, 20, 28

Challenging Days
3, 9, 15, 17, 23, 24, 25, 27, 29, 30

Scorpio/December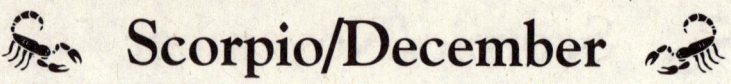

Planetary Hotspots
Mercury, the communication planet, switches direction in Capricorn, your solar Third House, on the 10th and retreats into Sagittarius, your solar Second House on the 18th. You can expect mix-ups and misunderstandings, as well as the possibility of dead batteries when Mercury is in Capricorn. Take extra precautions to safeguard financial information and to confirm all bills have been paid when it moves in Sagittarius. Postpone major purchases until later in January, or buy an extended warranty.

Planetary Lightspots
You sparkle with an extra level of magical, magnetic charisma, thanks to Venus in your sign all month. This annual influence also boosts your powers of attraction and your popularity quotient just in time for holiday socializing. Someone you meet the first week of December could be an invaluable contact in 2011.

Relationships
With Mercury retrograde, and Mars in Capricorn from the 7th on, you'll want to think carefully before you speak or send e-mail. A stressful moment could prompt a spontaneous response that you'll regret and be unable to retract. With that in mind, you're set for in-depth conversations with stimulating people. But steer clear of irritating neighbors and relatives, especially midmonth.

Money and Success
December is an upbeat financial month, with both the New Moon on the 5th and the Full Moon/lunar eclipse on the 21st in your money sectors. Be mindful of Mercury-related glitches, though. This duo could bring you, your mate, or both of you a nice year-end bonus and a surprise gift. Take a chance on the lottery the week of the Full Moon.

Rewarding Days
1, 2, 4, 8, 12, 13, 17, 25, 29, 30, 31

Challenging Days
7, 9, 14, 15, 20, 21, 24, 26, 27, 28

Scorpio Action Table

These dates reflect the best—but not the only—times for success and ease in these activities, according to your Sun sign.

	JAN	FEB	MAR	APR	MAY	JUN	JUL	AUG	SEP	OCT	NOV	DEC
Move	25, 26	7, 8, 12, 13										
Start a class	16–18, 27, 28	5, 6	9, 10						16, 17			
Join a club		1					15, 16	11, 12	14–21		1, 2	
Ask for a raise		8				24, 25		17, 18				
Look for work			16–31	1, 2								
Get pro advice				15, 16	13	7	6	2, 3				
Get a loan	25				14							
See a doctor			16–18, 25, 26									
Start a diet					10	6, 7	3–5, 31	1	23–25			
End relationship					12	8–10	6, 7		26, 27	23, 24	19, 20	
Buy clothes	18, 19	15, 28	1–6	10		4						12, 13
Get a makeover			4, 5		25				11, 12			2
New romance			15	10, 11		4	28, 29					
Vacation					25, 26	25–30	1–9		30	1, 2		

SAGITTARIUS

The Archer
November 21 to December 21

♐

Element: Fire

Quality: Mutable

Polarity: Yang/masculine

Planetary Ruler: Jupiter

Meditation: I can take time to explore my soul

Gemstone: Turquoise

Power Stones: Lapis lazuli, azurite, sodalite

Key Phrase: I understand

Glyph: Archer's arrow

Anatomy: Hips, thighs, sciatic nerve

Color: Royal blue, purple

Animal: Fleet-footed animals

Myths/Legends: Athena, Chiron

House: Ninth

Opposite Sign: Gemini

Flower: Narcissus

Key Word: Optimism

Your Strengths and Challenges

Enthusiasm and optimism are two of your greatest strengths, and you feel almost anything is possible if you truly believe in your mission. This idealistic viewpoint often leads to a positive outcome, but you can just as easily slip into wishful thinking. Learn to balance your hopes and wishes with reality.

Your adventuresome spirit makes you a modern-day explorer in search of truth, justice, and knowledge. Many Sagittarians love to travel and snap up any opportunity to see another part of the globe. Others are perpetual students or teachers, whether in the classroom or learning on their own. This also means your brain is a storehouse of information that can be useful in your career as well as everyday life.

But as a seeker and purveyor of truth, you also emphasize honesty—sometimes too much so. Telling it like it is isn't always the best choice, even though you have the best of intentions. A softer, more tactful approach can outweigh what you might perceive as duplicity and more readily achieve your aims.

Jupiter, your ruling planet, helps you see the big picture and reinforces your outgoing nature; it also encourages restlessness. This expansive planet is your lucky charm, and more often than not you land on your feet, ready for your next quest. Just don't depend upon Jupiter to come through for you all the time. Back it up with diligent effort.

Your Relationships

You excel at playing the field because independence is high on your priority list. That can make finding a lasting romantic relationship a challenge. But you're willing to trade at least some of your freedom for lasting commitment once you find your soul mate. That could be much later in life than your peers, and some Sagittarians never take the plunge. If you do, it's a must that your mate appreciate your spirit of adventure, which even a strong partnership won't change. You could fall in love with a Gemini, your opposite sign, and have much in common with the other fire signs, Aries and Leo. Life with a Libra or Aquarius would be mentally stimulating, but a clash is possible with Virgo or Pisces.

Your family is near and dear and you definitely have a soft spot for relatives. But you tend to idealize these relationships, and a parent may have been absent during your childhood years. For you, though, home is your haven; it's the place where you can retreat from the world and be most comfortable and secure. Emphasize consistency if you're a parent. Your children will benefit from that and your active participation in their lives and activities.

As independent as you are, you enjoy going places and doing things with friends rather than on your own. You're a wonderful friend to those closest to you, and you develop strong ties that can last a lifetime. In fact, commitment can grow out of what begins as a friendship because ultimately you want a partner who's both your mate and your best friend.

Your Career and Money

You're motivated by service to others and need a career that lets you provide what others want and need. When this desire isn't fulfilled you're quick to move on in search of the perfect, although nonexistent, outlet for your skills and talents. In the right career and job, however, you're content to stay put for years. Daily routine appeals to you, and as a hard worker you can be relentless in your determination to complete a project.

In money matters you know no in-between. You waver between thrift and extravagance and can live on a slim budget when necessary. Financial security becomes important to you in later years, when you're better able to achieve this goal. But be cautious about debt. It can be easy to get into a spending habit just to satisfy your need for emotional well-being. Teach yourself to save and pay cash.

Your Lighter Side

You're an armchair psychologist, interested in what motivates people and what makes them tick. This enhances your people skills, but more importantly it teaches you a lot about yourself. And that is information you can use to get ahead in the world and to match your outward confidence with the inner strength that can help keep you on track.

Affirmation for the Year

Home satisfies my adventuresome spirit—this year!

The Year Ahead for Sagittarius

You might surprise yourself this year—at least part of the time. With Jupiter, your ruler, in Pisces, your solar Fourth House, you'll undoubtedly enjoy more time at home. This makes 2010 a great year to get your place in shape, inside and out, and to host frequent gatherings for friends and family. Family relationships will be upbeat this year, and you'll want to spend time with relatives. Your home may in some ways become a "learning center" for self-study or online classes. Some parents may consider home-schooling, tutor children in academics, or teach music. On another level, this year is when you should create a foundation upon which to build your life during the next six years. At that time Jupiter will enter your solar Tenth House, when you can make great strides in your career and the world at large.

Jupiter briefly visits Aries, your solar Fifth House, June 5–September 7. These few months promise summer fun and romance. Besides, you'll be ready to get out of the house by then! As much as this is a social transit and one that attracts new dating relationships, it's also a creative one. So find a way to express that side of yourself, whether in an artistic endeavor or by learning a new hobby. Sports can also achieve the goal of taking your mind off the stresses and strains of daily life. Join a gym or sports league, or try tennis, golf, or another individual sport.

Your children will delight you, and you might want to get involved in their activities as a coach or group leader, or as a chaperone on a summer camping trip. This is a terrific time to become your children's #1 cheerleader, encouraging them to explore their skills and talents and giving them every opportunity to do so. If you're the parent of young children, their friends may spend so much time at your house that you'll feel like your family has multiplied—which it also could do with an addition to the family.

Saturn also travels in two signs, beginning and ending the year in Libra, with a brief and final return to Virgo, April 7–July 20. Placed in your solar Tenth House, Saturn in Virgo emphasizes your career, as it has done since it entered this sign in 2007. This is Saturn's strongest position, and you should take advantage of these few months to either firmly cement your career direction or to switch to something new. If you've earned success, Saturn will reward you. Be sure to follow the rules during this time and to be aboveboard—even if those around you are not quite so ethical. With Saturn here you also can be a mentor to others.

Saturn is in Libra, your solar Eleventh House, the rest of the year and through much of 2012. You'll undoubtedly make new friends during this time, including someone who's a soul mate or who enters your life to fulfill a specific purpose. Some friends from the past may reappear, and you'll make a conscious decision to distance yourself from others.

The Eleventh House is also associated with group activities, so don't be surprised if you have the urge to join a club or professional organization, or to get involved with a good cause. People will look up to you and expect you to take the lead. But they'll also expect you to do a major share of the work. Learn to delegate and fulfill only your responsibilities. Otherwise this will be a negative rather than a positive learning experience. The same applies to teamwork on the job, where you can excel at bringing others to consensus.

Uranus also switches signs this year, spending most of its time in Pisces, with a brief visit in Aries this summer as a preview for 2011. Uranus in Pisces emphasizes Jupiter's influence on your solar Fourth House, with one important difference: the need for domestic change. This can manifest as relocation, extensive remodeling, or someone moving in or out of your home. Family relationships are also likely to change in some way, mostly for the positive, and at least one of these people will bring you luck—an opportunity rather than an outright gift. So it makes sense to network with relatives, tell them what you need, and ask for their help. Return the favor.

The domestic scene will be lively this year, even more so than has been since Uranus entered Pisces. Expect unexpected visitors and family news, and plenty of home-based activities and people coming and going. Relatives and visiting friends can stimulate your thinking and spark sudden flashes of enlightenment. Listen closely. The answer you seek may be nothing more than a casual comment that helps everything click into place.

But Uranus in Pisces doesn't operate totally on its own this year. It will form an exact alignment with Saturn in Virgo in April, spanning your solar Fourth/Tenth House axis. This can trigger a career change for you or a family member. Downsizing is a possibility. Be cautious, however, if your thoughts turn to a home-based business. It may or may not be successful despite your hard work and optimism. This lineup can also prompt a sudden relocation as a result of career changes.

Uranus visits Aries, your solar Fifth House, May 27–August 12, a transit that can energize your social life and your love life. It's one of the best for a whirlwind romance (don't expect it to last), as well as playing the field. This is also your creative sector, so do something to in some way express your individuality, whether that's a hobby or fulfilling a long-held desire to climb a mountain or drive a race car.

Some Sagittarians will learn there's a new family member on the way, possibly an unexpected one, and others will see adult children move in or out, or go away to college. Be prepared to be firm if you have teenagers. They'll assert their independence more than ever before, so be sure to monitor their activities and, more importantly, get them involved in constructive ones, such as a part-time job or volunteer work.

As it does in Virgo/Pisces, Saturn and Uranus line up in Aries/Libra, across your solar Fifth/Eleventh House axis. Take the initiative to get acquainted with your children's friends and their parents; it's a wise move, especially if you have teens, as is helping them develop and live within a budget. The Saturn/Uranus alignment could also trigger a new friendship for you or rekindle one from the past. This relationship may or may not be in your best interests, so go slowly and don't feel obligated to loan money to any friend who asks. Also be cautious with investments; big gains and losses are equally possible.

You've probably experienced the many facets of Neptune since 1995, when it entered Aquarius, your solar Third House of communication. This mystical planet is known for confusion and illusion but also inspiration and spirituality. It can spark bright ideas, some on target and some not, and truly creative moments. With Neptune here you're also susceptible to hard-luck stories and seeing only the best in people. Unfortunately, not everyone shares your lofty ideals. Your enhanced intuition can be a real asset in these situations

In-depth discussions on many topics will hold your interest, especially those that center on the mysteries and meaning of life. Self-help books can inspire and motivate you, and provide solutions or at least aim you in the right direction regarding family and childhood issues, as well as how to maintain a positive mindset when tension gets the best of you.

Pluto, now firmly in Capricorn, will transit your solar Second House until 2024. A conservative financial approach is to your advantage now and in the coming years, and you should also take a close look at your attitudes and habits regarding money. This transformative planet can help you make necessary changes. Otherwise, Pluto may trigger events to achieve the same end. Save, invest for the long term, and do all you can to maximize income, because this transit can multiply your net worth as easily as it can leave you poorer but wiser. Also be sure your possessions and property are fully insured; this is not the time to gamble on the Universe's protection.

What This Year's Eclipses Mean for You

Pluto's financial spotlight is repeated in three of this year's eclipses—two in Capricorn and one in Cancer, your money sectors. The fourth eclipse is in Gemini, your solar Seventh House of relationships. Eclipse energy typically lasts about six months.

The January 15 solar eclipse in Capricorn, your solar Second House of personal resources, has great potential to increase your bank account, thanks to its alignment with Venus, also in Capricorn. Peak job performance could net you a nice raise or bonus. This eclipse also connects with Uranus in Pisces, so you could receive a gift or inheritance from a family member. With money flowing your way you might be tempted to seek a home improvement loan or purchase property. Be careful. This is not the year to over-extend yourself and to bet that income will continue to rise.

June 26 brings a lunar eclipse in Capricorn, but this one is not nearly as favorable. It aligns with Pluto in Capricorn and Mercury in Cancer, your solar Eighth House of joint resources. This combination will require some tough choices and decisions regarding money (your own and family funds). Loans could be easy to get but tough to repay, and you should check your credit reports for errors. This eclipse energy also has the potential to trigger a windfall because it contacts Jupiter and Uranus in Aries, your solar Fifth House. Take it easy, though. Jupiter can expand or shrink your net worth.

The July 11 solar eclipse in Cancer spotlights joint resources—family funds, debtors, insurance, and legacies. Its alignment with Mars in Virgo, your solar Tenth House, will give you and your mate the opportunity to increase income, although you'll need to take the initiative to make that happen. Also possible is a change in job benefits (more or less), and maintaining the same level could cost you more. You might also be successful in negotiating a promotion or more lucrative position, or earn a bonus for cost-saving idea for your company.

Gemini, your solar Seventh House, is the site of the December 21 lunar eclipse, which will focus your attention on relationships. Indecision regarding these ties will also be a factor because this eclipse aligns with retrograde Mercury in your sign. Listen to your inner voice if you're considering either a commitment or a breakup. Hold off, at least for a few months, if you're unsure. Because this eclipse also aligns with Jupiter and Uranus in Pisces, you could see a change in your home—a roommate, romantic interest, or family member moving in or out. Given a choice, however, this isn't the best time to invite someone new into your home. What begins with optimism is more likely to end with hard feelings.

Saturn

Saturn will contact your Sun from Virgo, your solar Tenth House of career and status, **if you were born between December 19 and 21**. You're among the last of your sign to experience this transit, and with it comes hard work and lowered vitality. Sleeping and eating well can help protect your immune system and maintain your high energy during this at times stressful period in your work life. Yet stellar achievements are also possible now (April 7–July 20) if you've lived up to Saturn's expectations. Saturn rewards effort and when in contact with your Sun always brings what you deserve.

Weigh the pros and cons and consider alternative options if you're questioning whether you're on the best career path for you. Also look at the big picture. It's possible you're in the right career but the wrong job. In any case, tread carefully. Career decisions made now will be in effect for the next fourteen to twenty-eight years. Patience might be your best asset now.

If you were born between November 21 and December 8, a favorable Saturn contact from Libra will be something of a breather. Saturn will twice contact your Sun (January 1–April 6, and July 20 through early August) **if your birthday is between November 21 and 26**; otherwise, Saturn's contact will be brief, a week or two, between mid-August and year's end. Transiting Libra, your solar Eleventh House, Saturn can bring endless, indecisive meetings, but also dynamic interaction with people and groups in both your personal and professional lives. Your challenge will be to keep others on track, to lead by example. Devote your time to only the most productive group activities—when you have a choice in the matter.

The Eleventh House also governs goals and objectives, hopes and wishes. This is your time to fully define what you want from life and what you want to accomplish, at least in the next few years. Security? Career gains? Successful relationships? Strong family ties? Friendship? In short, what will make you happy? Then tap into Saturn's talent for planning and organization and chart your short- and long-term paths. Doing this—getting yourself on track—is also likely to result in increased income.

Uranus

If you were born between December 15 and 21, Uranus will contact your Sun from Pisces, your solar Fourth House of home and family. Domestic changes will occur this year, mostly because you have the urge for change. Be sure that's what you want before you act because actions taken under a Uranus transit are almost always final, with little chance to reverse the decision. Consider some alternatives such as redecorating or remodeling your home and creating your own personal space within it—a hobby room or even a comfy chair in a corner reserved just for you. The point is that you need to do something to satisfy the strong freedom urge you have now.

But if Saturn in Virgo will also contact your Sun this year, career developments could trigger household changes. Downsizing could affect your job or your coworkers, resulting in extra responsibilities for you. Be alert to what's happening at your company the first three months of the year, and, if necessary, try to position yourself for the best possible outcome under the circumstances. If you want to sell your home (or you think it might be necessary), success is more likely early or late in the year. But a home purchase isn't the wisest choice now unless you're financially secure and plan to stay there for years to come; if so, you might find a terrific deal, especially on a property that could benefit from do-it-yourself skills.

Uranus in Aries, your solar Fifth House, will contact your Sun **if you were born November 21 or 22**. This favorable alignment with your Sun (May 27–August 12) gives you all the tools necessary to initiate positive, lasting change. You can remake yourself almost any way you choose now, whether your goal is weight loss, getting in shape, or switching to a healthier diet. Your thinking will be clear, even brilliant at times, and a flash of insight could encourage you to develop a latent talent. A chance encounter could trigger a fabulous romance, or you might suddenly discover that a friendship has much more potential.

Because Saturn will also contact your Sun, however, a friendship may suddenly end with no explanation, or you could unexpectedly connect with someone you haven't seen in years. Take note of these and similar events, for each comes with a lesson easily learned if you look for it. You may also decide to end your involvement with a group activity, or seek something new. This is more than likely to be short-term as your interest will wane when you discover it wasn't what you thought it would be.

Your children will surprise you during this time frame—hopefully for the better. With Uranus involved, however, you should stay in close touch with them and monitor their activities, Internet contacts, and friendships. Take swift action if necessary and view this as an opportunity to both teach and learn from your children. You'll certainly advance your knowledge of parenting skills.

Neptune

If you were born between December 15 and 21, Neptune will contact your Sun from Aquarius, your solar Third House. Your intuition will be active, helping you to sense what others are thinking. Listen when your inner voice speaks, and test it in small ways if this is a new experience for you. Creative thinking can help you in problem-solving as long as you're sure the ideas are both realistic and practical. Creative writing is another great use of this Neptune-Sun connection, so give it a try if this is something you've always wanted to do. Better yet, take a class that will teach you all the techniques you need to be successful. You might also ramp up your public-speaking skills because you'll have a way with words that can capture and hold an audience's attention.

Pluto

If you were born between November 25 and 27, money matters will be in the forefront as Pluto contacts your Sun from Capricorn, your solar Second House, and activates the June eclipse. Expenses may rise and income fall, so do all you can early in the year to build up savings. Then you'll have a reserve to rely on if necessary.

The Second House is also about possessions and what you value. Both will be prominent themes in your life this year and you'll be motivated to clean out junk, restore order, and eliminate items that are no longer useful. Sell the best at a yard sale or consignment shop, and donate the rest to a good cause. Both the process and the results will bring you great satisfaction and are a positive use of Pluto's energy. Just try not to get carried away and toss out something that might be a valuable collectible.

Jupiter-Saturn-Uranus-Pluto

If you were born between November 21 and 25, this four-planet alignment will contact your Sun from your Second House (Pluto in Capricorn), Fifth House (Jupiter and Uranus in Aries), and Eleventh House (Saturn in Libra). These involve money, creativity, children, gambling, romance, and leisure time, as well as friendship and groups. Almost anything associated with these areas is possible with Uranus involved and events will happen quickly. There could be extra expenses for your children, an organization might ask you to head up a fundraising campaign (think carefully!), someone could ask you for a loan, or a hobby or vacation trip could end up being far more expensive than you ever imagined. But you also could luck into a windfall or see one of your children excel in a major sports competition. There is one potential event this lineup cautions against: mixing money with romance or friendship. Chances are, you'll never again see your hard-earned cash.

Sagittarius/January

Planetary Hotspots
A friendship could end this month when values, opinions, or ideas clash. Or the same could happen with a club, organization, or group in which you're involved when everyone expects you to do all the work while they take the credit. You'll get the first inkling of what might happen around the 15th when Mercury resumes direct motion. Do what's right for you, but also give some thought to the karmic effects of potential actions.

Planetary Lightspots
You have an unusual year ahead of you. Noted for your adventuresome spirit, this year you'll really want to spend more time on the domestic scene. That's the effect of Jupiter, your ruling planet, in Pisces, your solar Fourth House. Enjoy! After all, it only happens once every twelve years. So make your to-do list and get started on home improvements.

Relationships
In-laws and other people may test your patience the week of the January 30 Full Moon in Leo, which activates retrograde Mars in the same sign. Patience will get you further, although it will be tough to keep your cool. Turn to your support system—friends, siblings, other relatives—for talk and stress relief.

Money and Success
Cross your fingers for a nice windfall; it could arrive in sync with the January 15 New Moon (solar eclipse) in Capricorn, your solar Second House. A raise or bonus is possible, as is a lucky find or a lucky win. You also can profit from selling unwanted items on consignment, and a thrift shop could yield a valuable collectible. At the least, you can cash in on some sensational sales.

Rewarding Days
1, 2, 3, 7, 10, 11, 15, 17, 21, 22, 25

Challenging Days
5, 6, 13, 16, 20, 27, 28, 31

Sagittarius/February

Planetary Hotspots
Life is fairly easygoing this month if you focus on any planet but Mars, which is still retrograde in Leo, your solar Ninth House. Like last month, it can trigger conflict, but it's more likely you'll simply feel frustrated when projects and plans stall. Consider it a lesson in patience, and remember that when you're behind the wheel, especially midmonth.

Planetary Lightspots
Venus enters your solar Fourth House on the 11th, making home life even more attractive. Since you're spending more time there, why not dust off your do-it-yourself skills and redecorate a room? You could net some nice bargains on household items around the 23rd, so watch for super-discount sales then. Aim for the last weekend of February if you want to host a get-together.

Relationships
You're at the center of the communication hub after Mercury enters Aquarius, your solar Third House, on the 10th. With the February 13 New Moon in the same sign, you'll have a way with words and be at your most charming at month's end, when you'll be able to convince just about anyone about just about anything.

Money and Success
Take time before the 10th to organize financial records and tax information. That way you'll get the benefit of Mercury in Capricorn, your solar Second House, and be able to check that chore off your to-do list. Your career comes into high focus on the 28th with the Full Moon in Virgo, your solar Tenth House. Put the lunar energy to work for you the following two weeks and snap up every opportunity to impress decision-makers. Actions taken now will pay off within a few months.

Rewarding Days
3, 6, 7, 8, 13, 14, 18, 26

Challenging Days
2, 5, 9, 12, 17, 20, 21, 23

Sagittarius/March

Planetary Hotspots
A dating relationship or friendship could end under this month's difficult planetary alignments in Aries, your solar Fifth House. Values could be the issue, or you might simply decide that it's far too confining for your lifestyle. If you're a parent, your children will need extra time and attention and you may discover just how expensive they can be. Also get to know their friends, and if you have teens, be prepared—they'll test your limits.

Planetary Lightspots
Home life is even more satisfying this month as lucky planetary lineups and the March 15 New Moon in Pisces, your solar Fourth House. You could luck into a great deal on a new home, but curb your optimistic enthusiasm when it comes to money. Rent less than you can afford and consider holding off on a purchase until this fall, when you might get an even better deal or decide you want to live elsewhere.

Relationships
The March 29 Full Moon in Libra shines brightly in your friendship sector. Despite the potential for tension with a pal, you'll delight in spending time with plenty of others and have the opportunity to welcome some new acquaintances into your circle. Think carefully, though, before you renew ties with someone from the past. Chances are, you'll wish you hadn't.

Money and Success
This is not the month for speculation, whether in the stock market or in a game of chance. The potential for loss far outweighs the possibility of gain. The same applies to anyone (even a family member) who asks for a loan or pressures you to cosign for one. Protect your resources.

Rewarding Days
6, 7, 12, 13, 14, 15, 17, 21, 22, 26, 30

Challenging Days
1, 2, 8, 11, 16, 20, 23, 25, 29

Sagittarius/April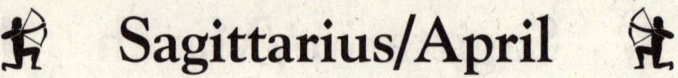

Planetary Hotspots
A career change could be on the horizon. Saturn returns to Virgo, your solar Tenth House, where it clashes with Uranus in Pisces. Downsizing is as possible as a step up, so you'll want to be alert to what's happening in your environment and to position yourself for a positive outcome. Relocation is also possible, but this is not the time to purchase property or to welcome a roommate.

Planetary Lightspots
Take a break around the time of the April 28 Full Moon in Scorpio, your solar Twelfth House. Even if only for a day, you'll benefit from time alone away from the hustle and bustle of daily life. The lunar energy also activates your sixth sense and quiet time will encourage your inner voice to emerge with insightful messages.

Relationships
Your social life gets a boost from the April 14 New Moon in Aries, your solar Fifth House. This is also a real plus for romance, a dating relationship, creativity, and quality time with your children. Take it a step further and get involved in your children's sports activities or join a gym. Meet your neighbors if you're single. Among them could be someone who catches your romantic interest.

Money and Success
Despite what's going on in your career sector, this month brings positive planetary alignments in Taurus, your solar Sixth House of daily work. That's positive for advancement, and if you're job hunting, put networking high on your priority list. But there is a potential downside here: Mercury turns retrograde in Taurus on the 17th, which can cause delays and indecision. Be patient.

Rewarding Days
2, 3, 8, 9, 13, 14, 17, 21, 26, 27, 30

Challenging Days
4, 5, 7, 12, 15, 19, 22, 23, 25, 28

Sagittarius/May

Planetary Hotspots
This month's Jupiter/Saturn alignment across your solar Fourth/Tenth House axis can make it tough to balance career and family responsibilities. The biggest challenge is that you're more interested in the domestic scene than you are in your career, at least early in May. And that could prompt you to think about a home-based business. Think again. This is not the time to invest in an entrepreneurial venture, and with a little patience the urge will pass.

Planetary Lightspots
You're set for new adventures as the May 27 Full Moon in Sagittarius lights up your life. Slow down and take a step back before you dash into the future. Think about what you've accomplished in the past six months and what you want to achieve between now and year's end. Then put your plans in writing and follow the path.

Relationships
Most of the relationships in your life benefit from Venus in Gemini, your solar Seventh House, through the 18th. Personal ties are at their best, but you could have some challenges with people at work, especially the boss. Take a different view. This is someone you can learn from and who can ultimately become one of your biggest supporters. Listen more than you talk and take the advice and constructive comments to heart.

Money and Success
Just as you question your career, you get a jump start from the May 13 New Moon in Taurus, your Sixth House of daily work. Even better, the odds favor a raise or bonus for your effort. So stick with it, work through the challenges and celebrate your success.

Rewarding Days
1, 5, 6, 10, 11, 14, 15, 19, 24, 28

Challenging Days
2, 4, 7, 8, 9, 16, 18, 22, 23, 29

Sagittarius/June

Planetary Hotspots
Month's end brings financial tension when the June 26 Full Moon in Capricorn, your solar Second House, activates Pluto, also in Capricorn. Your or your partner's income could decrease, or you may have a major expense, possibly related to one of your children. Investments require maximum caution now, and what seems like a safe bet could backfire and leave you poorer but wiser.

Planetary Lightspots
Your social life gets a boost from Jupiter, which enters Aries on the 6th. Two days later it merges with Uranus in the same sign, highlighting a week that could feature a fabulously lucky chance encounter. You could connect with a terrific career contact, or even meet someone who changes your life. For singles, this planetary duo could trigger a love-at-first-sight romance; if you're among them, however, take things slowly because it could end as quickly as it begins.

Relationships
Relationships are mostly upbeat this month, especially the week of the June 12 New Moon in Gemini, your solar Seventh House. Communication flows with your partner, relatives, and neighbors, but a workplace relationship could be strained. Go with the boss's wishes even if you disagree or know for a fact that you have the right idea.

Money and Success
Although you'll find coworkers especially supportive as Mercury transits Taurus, your solar Sixth House, through the 9th, one of them could have a hidden agenda. It's also well worth listening to the grapevine as long as you don't take everything you hear as fact.

Rewarding Days
1, 7, 8, 11, 14, 15, 16, 20, 28, 29, 30

Challenging Days
3, 4, 5, 6, 10, 12, 19, 24, 25, 26

Sagittarius/July

Planetary Hotspots

July brings another Uranus-Saturn alignment, this time in Aries-Libra, your solar Fifth/Eleventh House axis. A friendship or dating relationship could end under this stressful lineup, or you might decide to cut ties with a group when you realize it's more restrictive than productive. If you're a parent, you'll want to monitor your children's activities and friendships, and it would be wise to get to know their friends' parents. Also be especially cautious with investments and speculation; a big loss is more likely than a big win.

Planetary Lightspots

You'll be an attention-getter from the 10th on, when Venus is in Virgo, your solar Tenth House. This planetary placement is an asset for your career, and can elevate your status in the world at large. So plan ahead to make the most of this beneficial influence because it only occurs once a year.

Relationships

The July 25 Full Moon in Aquarius, your solar Third House, emphasizes communication, as does Mercury in Leo, your solar Ninth House, from the 9th to the 26th. You'll have a way with charming words now and people will look to you for information and knowledge. Share what you know, teaching others (formally or informally) with patience and insight.

Money and Success

July features a New Moon (solar eclipse) in Cancer, your solar Eighth House. This positive financial influence can boost your bank account during the next six months. But it also requires wise stewardship of family funds. Be thrifty and save. A raise is possible within the first two weeks of July for you, your mate, or both of you.

Rewarding Days

4, 5, 8, 12, 13, 17, 26, 27, 31

Challenging Days

3, 9, 10, 16, 21, 22, 23, 28, 29, 30

Sagittarius/August

Planetary Hotspots

Tension surrounding your solar Fifth and Eleventh Houses continues this month, and with several planets in those sectors clashing with Pluto in Capricorn, your solar Third House, you can expect recent issues to reach a peak. Money may be involved, so caution is the only option with investments. In addition, don't let anyone pressure you into a loan (including cosigning for one) or donation, even if that means you cut ties. Protect your financial security.

Planetary Lightspots

This month's New Moon in Leo on the 9th is designed with you in mind. Placed in Leo, your solar Ninth House of travel and adventure, you're ready to add some excitement to your life. Time permitting, dash off to parts known or unknown for a vacation or long weekend. Either one will satisfy your restless feeling and provide a much-needed change of scenery.

Relationships

Romance, recreation, social events, and children capture your attention under the August 24 Full Moon in Pisces, your solar Fifth House. And with Uranus returning to Pisces on the 13th, you'll be in a spontaneous mood. Call friends with spur-of-the-moment invitations, surprise your mate or a date, and indulge your children's requests when they least expect it. And don't forget yourself; express your individuality through a favorite leisure-time activity.

Money and Success

Focus on the details from the 20th on, the date Mercury turns retrograde in Virgo, your solar Tenth House of career. Errors can easily slip through. Also double-check deadlines and instructions, and think before you speak during this time that's prone to mix-ups and misunderstandings.

Rewarding Days

1, 5, 8, 9, 13, 14, 17, 21, 22, 23

Challenging Days

4, 6, 10, 12, 18, 19, 20, 25, 27, 28

Sagittarius/September

Planetary Hotspots
Although issues associated with this summer's events begin to wane, the Full Moon in Aries on the 23rd will activate both Saturn in Libra and Pluto in Capricorn. This influence continues to urge caution with money matters, especially those that involve children, speculation, friends, and groups. You also should give serious thought to what you value, what's most important in your life. On a practical level, this is a great lineup for tossing out junk. In the process you could find a valuable collectible.

Planetary Lightspots
Jupiter returns to Pisces, your solar Fourth House, on the 8th, where it joins forces with Uranus on the 18th. This lineup is a plus for family relationships and relaxing yet stimulating get-togethers with friends at home. But be careful with electricity, check appliances regularly, and be sure to back up computer files.

Relationships
Despite the stress and strain associated with Saturn in Libra, your solar Eleventh House, you'll also have special moments with close friends while Venus transits that sign through the 7th. Get together with your best pal early in the month when planetary alignments will enhance communication. A long talk will inspire you and renew your faith in yourself and the Universe. Later in the month, however, do all you can to avoid difficult people.

Money and Success
The September 8 New Moon in Virgo highlights your solar Tenth House, boosting your potential for career gains. It also could trigger a surprising development that you can work to your advantage. Listen to the grapevine in the days before the 12th, when Mercury resumes direct motion in Virgo.

Rewarding Days
2, 5, 6, 8, 10, 11, 14, 17, 20, 24, 28

Challenging Days
3, 9, 13, 15, 16, 18, 19, 23, 25, 26, 30

Sagittarius/October

Planetary Hotspots

Venus turns retrograde on the 8th, so don't be surprised if you covet more time alone, out of the social circle. Take it. This is a great time to catch up on sleep, devote time to your favorite activities and generally re-center yourself before the Sun enters your sign. If you have a pet, now would be a good time to schedule a checkup for your furry friend. Do the same for you.

Planetary Lightspots

Your sixth sense is unusually active this month with several planets transiting Scorpio, your solar Twelfth House. Tune in, observe, and listen to the grapevine. You can pick up a lot of valuable information as well as personal insights that will help you resolve what's on your mind.

Relationships

The October 22 Full Moon in Aries, your solar Fifth House, is as great for singles looking for love as it is for couples celebrating togetherness. Plan a special evening for two, or take a chance on someone who catches your eye. If you're a parent, you could meet someone through your children, who will appreciate your extra time and attention now. Consider getting involved in their sporting or after-school activities as a coach or fund-raiser.

Money and Success

Much of this month's career activity will be behind the scenes, and may include confidential information that you'll ultimately be able to profit from. This is definitely a situation where who you know is more important than what you know, so get acquainted with the right people and show them you have what it takes. With luck, which you have now, you might even gain financially.

Rewarding Days

7, 8, 10, 11, 12, 15, 16, 17, 25, 26

Challenging Days

2, 3, 4, 6, 13, 20, 27, 28

Sagittarius/November

Planetary Hotspots
You're unusually accident-prone with Mars in your sign all month. This calls for caution on the road, in the kitchen, working with tools, and even walking and going about your daily activities. Haste is the culprit, so slow down and curb your impatience. Do the same with other people, because Mars can trigger conflict even if that's not your intention. Soften your words.

Planetary Lightspots
The New Moon in Scorpio, your solar Twelfth House, on the 6th urges you to continue to take it easy, at least early in the month. Rest, sleep, and a healthy diet will also help prevent a cold or flu. Take it one step further and pamper yourself with a massage or the full treatment at a day spa. You deserve it.

Relationships
Venus resumes direct motion on the 18th in Libra, your solar Eleventh House. The timing is ideal for holiday social events with friends, meeting people, and developing new networking contacts. Line up outings, and consider hosting a get-together. With Jupiter in Pisces, your solar Fourth House, also turning direct on the 18th, you can quickly get your place in shape for an open house or Thanksgiving feast with friends and family.

Money and Success
Your workload will pick up the week of the November 21 Full Moon in Taurus, your solar Sixth House, and it will take organization and planning to complete a lengthy to-do list. But you'll also reap great satisfaction from the effort and earn well-deserved praise. Strive to complete everything you can because December will be hectic.

Rewarding Days
4, 6, 7, 8, 12, 13, 16, 18, 22, 26

Challenging Days
3, 9, 15, 17, 23, 24, 25, 27, 29, 30

Sagittarius/December

Planetary Hotspots
December's hotspot is Mercury, which turns retrograde in Capricorn, your solar Second House, on the 10th. That can trigger mix-ups with bill payments and accounts, and it also advises against the purchase of electronics or appliances. If either is must, check the return policy and consider an extended warranty. That's not all, however. Mercury retreats into your sign on the 18th, which can interfere with personal plans. Look forward to the 30th, when Mercury turns direct.

Planetary Lightspots
December 5 brings the New Moon in your sign. That's plenty to cheer about, because you'll be the center of attention more often than not at social events and just about anywhere you go. Also give some thought to what you want to accomplish personally and professionally in the year ahead and set plans in motion when Mercury resumes direct motion.

Relationships
Relationships are all you could wish for with the December 21 Full Moon (lunar eclipse) in Gemini, your solar Seventh House. For some, the lunar energy could trigger a new romance or engagement, but most of all you'll delight in being with people, especially loved ones and close friends. But don't be hasty if you're considering a new home or sharing yours with a roommate or romantic interest.

Money and Success
Despite retrograde Mercury, you could be in line for a nice year-end bonus, thanks to Mars in Capricorn from the 7th on. The red planet could also trigger an opportunity to earn more this month. Save rather than spend if you can, and stick to a thrifty gift budget. There could be extra expenses this month.

Rewarding Days
1, 4, 5, 10, 15, 19, 23, 24, 31

Challenging Days
6, 11, 14, 20, 21, 26, 27, 28

Sagittarius Action Table

These dates reflect the best—but not the only—times for success and ease in these activities, according to your Sun sign.

	JAN	FEB	MAR	APR	MAY	JUN	JUL	AUG	SEP	OCT	NOV	DEC
Move	19–28	1–6				4, 5			21, 22			
Start a class	20–31	10–17								15, 16		
Join a club						19, 20		13, 14		4–19		
Ask for a raise	13, 14			3–16					16, 17			7
Look for work					12–19	8		2, 3				
Get pro advice	25, 26				14	10–18						
Get a loan		24, 25			25, 26		10		30			
See a doctor			4, 5	15		8		2, 3				2
Start a diet						8	6, 7	5, 6	26, 27		19, 20	
End relationship									1, 2, 28, 29		22, 23	
Buy clothes			16–31		10							
Get a makeover		8				24, 25	22	17, 18				
New romance			16, 17	12–14								
Vacation			25, 26				23–26	9, 10				

CAPRICORN

The Goat
December 21 to January 20
♑

Element: Earth
Quality: Cardinal
Polarity: Yin/feminine
Planetary Ruler: Saturn
Meditation: I know the strength of my soul
Gemstone: Garnet
Power Stones: Peridot, diamond, quartz, black obsidian, onyx
Key Phrase: I use
Glyph: Head of goat

Anatomy: Skeleton, knees, skin
Color: Black, forest green
Animal: Goats, thick-shelled animals
Myths/Legends: Chronos, Vesta, Pan
House: Tenth
Opposite Sign: Cancer
Flower: Carnation
Key Word: Ambitious

Your Strengths and Challenges

Born under the most ambitious sign of the zodiac, you push yourself to achieve. Long-term planning is central to your success and you weigh major decisions in terms of their potential impact on the future. Experience is another important tool, and you use this knowledge to help build a solid foundation. With Saturn as your ruling planet you can be overly cautious. And although that's to your advantage in some situations, in others a well-calculated risk could better advance your aims. Expand your comfort zone in small ways so you're prepared to extend yourself when the opportunity arises.

Organized, patient, and methodical, you're at times practical to a fault. Yet these qualities, along with your strong sense of responsibility, yield a conscientious approach throughout your life. You're in part motivated by a fear of failure, which, although usually unfounded, is a driving force in your quest for status and success.

Most Capricorns begin to reap rewards for their efforts around age forty, having gained the wisdom they need to excel and outdistance the competition. Life before then is a training ground, so use it wisely. Despite your focus on success and achievement, you can benefit from leisure time—even if it too has a practical benefit. Strive for balance by developing a hobby such as gardening or woodworking, and set aside weekend time for family and friends. You'll be all the better prepared to tackle your goals!

Your Relationships

You have an unbreakable bond with your closest friends, who are few in number. These deep emotional connections are part of the foundation of your life, and each serves a purpose. Yet these friends do not necessarily know—or know of—each other. Among your many acquaintances are powerful people you meet through your career, and you excel at networking with them. It's part of your secret to success and climbing the ladder.

You maintain close ties with your family because you have an inseparable, if often unspoken, connection with them, particularly your mother. Yet you also follow through with your desire to have a life of your own. As a parent you're affectionate and fond of your children. You also have a no-nonsense attitude and set a firm but fair level of expectations and discipline while providing them with every opportunity to learn and grow as individuals. One of your children may push your limits more often than not. If so, take a close look. Chances are, you're very alike, and you can form the strongest of bonds with this child.

Most outsiders are unaware of your tender heart. They see only the ambitious Capricorn who's all business with little time for the softer side of life. But you're ultimately passionate and sensual with the right person. Possessive in dating relationships (and unwilling to let go even when you know you should), you're less so in a partnership because commitment represents security. A Scorpio or Pisces might be your ideal partner, and you're in sync with fellow earth signs, Taurus and Virgo, as well as other Capricorns. Your best match could be a Cancer, your opposite sign. But you'll want to be cautious about linking your heart with an Aries or Libra, despite the sizzle.

Your Career and Money
Career and ambition define you. You also plan every step and usually (but not always) have the patience to wait until the time is right to make your next move. So it's tough to hold yourself back when you're ready to move on and up, even if you know that's the best option. In your daily work you need a lot of variety and communication, both of which offset your low threshold for boredom. Many Capricorns thrive in a corporate environment.

In money matters you're cautious, conservative, and thrifty; but some Capricorns take this too far and become miserly. Investments are usually profitable over the long term because you base decisions on facts and figures rather than emotions. Your financial weakness is big-ticket items that are symbolic of your success—a big home, the latest technology and electronic gadgets, and a status vehicle.

Your Lighter Side
Your inner being is frivolous to the max, although most people never see this side of you. Enjoy! And let this side of yourself emerge more often. Being carefree and irresponsible (within Capricorn limits!) is the one side of the real you that wants to experience ever facet of life to its fullest. Play! Smile! Laugh!

Affirmation for the Year
Determination and flexibility are my best career assets.

The Year Ahead for Capricorn

Jupiter enters Pisces, your solar Third House, January 17, and the lucky planet will spend most of the year there, with a brief visit to Aries this summer. During Jupiter's time in Pisces, you'll be on the go between errands, chauffeuring children if you're a parent, and dashing off for the weekend. The Third House is also your communication sector, so expect to be in touch with just about everyone you know at some point during 2010. Keeping up with calls and e-mail will test your organizational skills at times.

A relative, especially a sibling, could bring you luck at the least expected moment. A casual comment might be just the link you need to a great job or another opportunity that shines brightly toward the future. You'll enjoy get-togethers with extended family and neighbors (an excellent networking source) and could be the one who launches a campaign to better your community. If you've ever wanted to run for city council or another local position, you could be a winner this year. Most of all, your thinking will be upbeat and optimistic, and your mind will absorb all the information it possibly can.

Jupiter's brief visit to Aries, your solar Fourth House, May 27–August 12, promises an active family life with plenty of interest in the domestic scene. Use this period for home improvements and routine maintenance, and devote a few weekends to cleaning closets and storage spaces and generally getting your place in order. You'll have the incentive and initiative to zip through it all in record time. Home will be your favorite location this summer, a place to relax and unwind from career-related responsibilities and tension. Consider taking a vacation at home, which you'll probably enjoy far more than a trip. Also take advantage of this time to reconnect with extended family—maybe even plan a reunion.

Saturn returns to Virgo this spring for its final transit of your solar Ninth House, April 7–July 20. This is the time to complete unfinished business begun since Saturn entered Virgo in 2007 (or even what you started or hoped to do seven or fourteen years ago). That could be finishing a degree or certification necessary to fulfill career ambitions, or a major trip you've postponed. Business travel will keep some Capricorns away from home, and others will find themselves in close contact with in-laws during these months. It's likely this transit is also related to career events that have occurred since late last fall. Look for a prevailing theme or repeated message and also listen to your instincts. Chances are, there is something you can do now to reinforce or advance your worldly aims.

You might also resolve a long-standing philosophical dilemma, or return to or reject a religious affiliation. In that sense Saturn challenges you to examine your belief system and make necessary changes; in other words, be true to yourself and stand firm with your personal code of ethics.

For an ambitious Capricorn, life is at its best when Saturn, your ruling planet, is in your solar Tenth House of career. The best news is it's just getting started, having entered Libra late last year. The next few years are your chance to shine. If you're like most Capricorns you're ready for the challenge simply because you've awaited this opportunity and planned for it—possibly for many years. Rising to the top comes naturally to you and you often have a sixth sense about when to make your move. This year, however, may not be the best choice despite a strong desire to shift gears.

Saturn in Libra clashes with Pluto in Capricorn, which could be very frustrating, both personally and career-wise. Power plays are possible with this planetary duo in action, so try not to get drawn in; let others deal with their own issues. You also could have difficulties with a supervisor; here, again, the wise choice is to keep a low profile, follow the rules, and fulfill your responsibilities. Also be careful whom you trust; an apparently friendly coworker could try to undermine your position.

But this transit can also launch you on the fast track to an influential position. Just be aware that it may take until later in the year for everything to come together. Trying to push events rather than letting things unfold in their own time could backfire.

Uranus also transits two signs this year: Pisces and Aries. By now you're fairly familiar with the effects of Uranus in Pisces, your solar Third House, since it entered that sign in 2003. This year promises more of the same, although to the nth degree because of Jupiter in the same sign. What Uranus adds is the element of the unexpected—sudden news and events. Some of it will concern a neighbor or relative, and you'll undoubtedly have flashes of insight when your intuition is active. Pay attention; this information could put you in the right place at the right time to snap up an opportunity. You also could suddenly achieve celebrity status in your community.

This transit encourages learning and will trigger your curiosity in a wide range of subjects. Reading, television, the Internet, puzzles, games, and talking with people will capture your interest. Uranus, like Saturn in Virgo, encourages you to further your education as a step toward career advancement. This is especially true because Saturn and Uranus form their final Virgo/Pisces alignment in April. The same lineup could trigger a legal matter or bring one to conclusion, probably with an unexpected outcome. Drive with extra care this year, and always socialize with a designated driver.

Emphasis shifts to your solar Fourth House of home and family when Uranus visits Aries, May 27–August 12. Changes on the domestic scene are the norm with Uranus here, although they may not occur in your life this year. Nevertheless, change or the desire for it will occur on some level in 2010, even if it's only an idea that drifts through your mind. One of the best uses of this transit is home improvements, beginning with the clutter that surrounds you. Relocation is likely either this year or in coming years, possibly several times. (Uranus spends approximately seven years in a sign.)

Be sure your property and possessions are adequately covered by insurance. Damage could result from severe weather or an appliance. Be especially cautious with electricity (forget do-it-yourself!), and always back up computer files. Even the least-expected event is to be expected when Uranus is on the scene, including people dropping in unannounced.

Neptune continues its long transit of Aquarius, your solar Second House, which it entered in 1995. You've undoubtedly experienced the many facets of Neptune's influence on money matters, including ups and downs, intuitive purchases, lost items, and creative ideas to enhance your bottom line. As much as Neptune can have you feeling disillusioned about your earning power, it also has a reputation for providing exactly what you need when you need it. In fact, this transit can ultimately increase your net worth by the time it completes its Aquarius transit. But you'll need to continue to be vigilant about finances. Steer clear of anything that sounds too good to be true; chances are, it is.

Neptune in the Second House is also about values and what is important to you, both personally and materialistically. Valuing yourself should be your first goal, and you should also think about your attitudes toward things. The more closely you identify with material objects, the greater the chance you will lose some of them. Let your ego speak through your accomplishments rather than what you can purchase with them.

Pluto will be in Capricorn until 2024, so you may not fully experience its effects until it contacts your Sun. Even so, this powerful planet will impact your life in some way this year, if only through events in your immediate environment. Observe what happens around you and how Pluto influences other people and the world at large. You will glean valuable knowledge for the future when Pluto motivates you to travel the path of personal transformation. Tune in to Pluto if you feel an increasing need for change on any level. Do remember, though, that actions taken under a Pluto transit are rarely reversed. But this is also Pluto's way: re-creating yourself to meet the future.

What This Year's Eclipses Mean for You

This year's eclipses are all about you, your relationships, and your work life, with the emphasis on you. For that you can thank two eclipses—one lunar and one solar—in your sign. An eclipse is active for approximately six months. The year gets off to the right start with the January 15 solar eclipse in Capricorn. This eclipse energy merges with Venus, also in Capricorn, and favorably contacts Uranus in Pisces, your solar Third House. That's a beautiful combination for romance, and a chance encounter will bring a new love interest. The reason is simple: you'll have an extra level of irresistible pizzazz, and more importantly, the eclipse will boost your powers of attraction. Expect plenty of attention and popularity, and use this lineup to attract exactly what you wish for in the coming months.

The June 26 lunar eclipse in Capricorn may be a bit more stressful, however. It's aligned with Pluto in Capricorn, Mercury in Cancer (your solar Seventh House), and Jupiter and Uranus in Aries (your solar Fourth House). This eclipse could also spark a whirlwind romance, or prompt you to question an existing business or personal relationship. Communication will be as much the solution as the problem, so both you and the other person will need to commit to open minds and compromise. Issues may be related to home and family, especially relatives.

The eclipse energy shifts to Cancer, your solar Seventh House, with the July 11 solar eclipse in that sign. Expect to be in contact with many people. Some Capricorns will commit to a partnership, and others will become involved in a long-distance relationship, possibly by reconnecting with someone at a reunion. Overall, people will be supportive of your aims and be more than willing to help out when needed. Build good karma and return the favor.

December 21 brings a lunar eclipse in Gemini, your solar Sixth House of work and wellness. Even if you're more or less content with your job, you'll feel somewhat out of step at times, and feel discontent. Making a decision won't be easy now, so think of this period as time in which you can examine what exactly you want from a job. Give meditation a try. Also find your best avenue to minimize stress, which will help your immune system prevent a cold or flu.

Saturn

If you were born between January 17 and 20, Saturn in Virgo, your solar Ninth House, will contact your Sun between April 7 and July 20. This favorable connection from your ruling planet is a steadying influence, giving you extra stamina and determination. Knowledge is a major theme with this transit, and you should open your mind to new ideas and viewpoints, including learning more about yourself. Self-knowledge is, after all, a great ally, so delve into the inner recesses of your mind through meditation or self-help books. You'll also have the patience now to work through tedious tasks and to tolerate people who aren't as quick as you are.

Business or personal travel could keep you away from home, and you may have to deal with a legal matter, possibly in connection with a relative. You can expect delays in these areas as well as unexpected events, especially in April when Saturn in Virgo will align with Uranus in Pisces. Even though both these planets will favorably contact your Sun, it's a wise idea to be cautious in communication; say and write less rather than more, and keep confidential information to yourself. What you learn now will come in handy in the near future.

If you were born between December 21 and January 6, your career is in focus this year as Saturn contacts your Sun from Libra, your solar Tenth House. But **only Capricorns born between December 21 and 26** will experience the main thrust of this Saturn transit, not once but twice—January 1–April 6, and July 21 through early August. Other Capricorns will feel it for a few weeks between mid-August and year's end. This Saturn-Sun contact can be difficult or rewarding, or a combination of both. Much depends upon the career (and career-related) decisions and actions you've taken during the past fourteen years. Transiting Saturn rewards or denies, and in almost every instance will bring what you've earned and deserve.

Saturn's transit this year is not so simplistic, however, because it will clash with Pluto in Capricorn in January and August. Reflect on what career events occurred last November when these two planets first aligned. This will give you some insight into what to expect from Saturn this year. Global conditions could be a strong indicator, or your company may undergo a significant restructuring. This probably isn't the best year to switch jobs unless it's absolutely necessary. What occurs in one job is likely to repeat at the next.

If your job is secure but your workplace environment is difficult, avoid controlling people by working around them. On the other hand, careful moves and contact with the right people now could lay the groundwork for a major step up later this year or next. Caution! It will be tough to know who is a true supporter and who is a "friendly" foe.

Uranus
If you were born between January 13 and 18, Uranus will contact your Sun from Pisces, your solar Third House of communication, quick trips, and learning. This favorable connection is ideal for on-target thinking, bright ideas, and sudden insights. Listen to your intuition; it will be especially active this year. With Jupiter also in Pisces much of the year, lucky moments can pop up just about anytime, so you'll want to meet new people and network at every opportunity. You'll also benefit from hidden support—people who cheer you on without your knowledge. Volunteer activities and charitable causes may appeal to you this year even if never before. Do something to better your community. Since you'll be on the go more than usual, be sure your vehicle is fully covered by insur-

ance. An extended warranty might be a good choice for anything mechanical.

Learning is a strong theme this year, whether on a casual basis or in a formal setting. Check out online classes. You should consider higher education or advanced or specialized training **if your birthday is between January 17 and 20**. Although studying will stretch you thin at times, it would be an excellent investment in your future.

If you were born December 21 or 22, change is on the horizon with Uranus contacting your Sun from Aries, your solar Fourth House, May 27–August 12. This can be one of the most exciting periods of your life, but there's also a potential downside with this transit. Stop and think things through if you feel uncharacteristically compelled to turn your life upside down. You may soon regret the decision. A good alternative is to update your image or your home. With Uranus in your solar Fourth House of home and family, relocation is possible, you might undertake an extensive remodeling project, or someone could move in or out. Be very cautious, though, if you want to purchase a home or launch a home-based business. Either one could become an expensive mistake. Update your property (or renter's) insurance to cover any eventuality, don't postpone repairs, and do all you can to minimize risk.

Neptune

If you were born between January 15 and 19, money matters will require more of your attention and care this year as Neptune contacts your Sun from Aquarius, your solar Second House of personal resources. Even though your thrifty nature will somewhat offset this influence, money will still have a tendency to slip through your fingers and disappear as if into thin air. But it can just as easily appear out of nowhere just when you need it. Possessions will be lost easily now, so take precautions when you're out in public, even in the safest of environments. Your financial thinking may seem as though it's on target, but that won't always be true this year. Avoid any major decisions if you can. When that's impossible, seek another opinion. In any case you should take no risks. Question any advice you receive. Good intentions and promises are not a guarantee.

Pluto

If you were born between December 24 and 27, powerful Pluto in Capricorn will merge its energy with your Sun this year. Your determination and willpower will know no bounds, and you'll work hard to achieve whatever tasks you set out to accomplish. But in doing that you can get too close to what you're involved in and lose perspective. Pluto is, after all, the planet of obsession.

Pluto takes change to the level of transformation, removing the old and replacing it with something new. You and your life should be the main focus here, so decide within the first few weeks of 2010 exactly what it is about yourself to want to change. Resisting Pluto is not an option, as you'll soon realize, so it's far better to initiate the action yourself. Only when this transit is in the past will you fully realize the depth of the changes—and how positive they are—as you greet your newly transformed self.

Jupiter-Saturn-Uranus-Pluto

If you were born between December 21 and 24, this four-planet alignment will contact your Sun this summer from your solar First House (Pluto in Capricorn), Fourth House (Jupiter and Uranus in Aries), and Tenth House (Saturn in Libra). This unusual and potentially stressful lineup signifies major change involving you, your home and family, and your career. Although it's difficult to second-guess any lineup that includes Uranus, there are possibilities, and you can be sure that events will unfold rapidly.

Career changes are possible, even likely. You may relocate for a job or downsizing could occur at your company. Be proactive and do all you can to protect your position even if it means taking on more responsibilities. You also could be offered a terrific position; but be cautious, because what begins in a flash could be just as short-lived.

A family member may require significant assistance. Look at every other option, however, before inviting a relative to join your household as this would be far more disruptive than you imagine. The same is true of a potential roommate. You should postpone a property purchase until next year if possible. It also will be difficult to sell your home at this time, so plan ahead and try to sell earlier in the year. Consider replacement insurance for your home and property, especially if you live in an area that's high risk for severe weather.

Capricorn/January

Planetary Hotspots
Saturn in Libra, your solar Tenth House, squares off with Pluto in Capricorn this month. Pay close attention to what's happening on the career front. Listen to rumors. Observe supervisors. You may pick up a hint of what will occur within a few months, although the first developments may become apparent the third week of January. Take care. Keep your distance if you sense conflict brewing.

Planetary Lightspots
You are this month's lightspot! For this you can thank the January 15 New Moon (solar eclipse) in your sign. Even better, the New Moon is aligned with Venus, also in your sign, multiplying your powers of attraction. Share your wishes with the universe because almost anything you want can be yours in the coming months. The secret? Believe you deserve it!

Relationships
Jupiter enters Pisces, your solar Third House of communication, on the 17th. This year-long influence is a real plus for learning and charming. You'll have a way with words and people will be especially receptive to your requests. Do yourself a favor. Check out fun classes, or take a few short-term ones to enhance your career skills.

Money and Success
You may feel a slight financial pinch the next few months as Mars travels retrograde in Leo, your solar Eighth House. So it will pay to be thrifty and save. Unexpected expenses could arise at month's end, when you might also have the urge to splurge. However, you could get a terrific deal on career clothing if you shop sales and use discount coupons the third week of January.

Rewarding Days
1, 2, 4, 9, 14, 17, 18, 19, 28

Challenging Days
3, 5, 6, 12, 13, 20, 23, 27, 31

Capricorn/February

Planetary Hotspots
You have a terrific month ahead for the most part, with only minor hurdles and challenges, and much success potential. All except for Mars, which is still retrograde in Leo, your solar Eighth House. Be cautious with investments and take time to review insurance policies to be sure your home, property, and possessions are fully covered. Pay premiums early, and if there's a problem with a claim, wait until after the 13th to challenge it.

Planetary Lightspots
Plan a weekend getaway around the time of the February 28 Full Moon in Virgo, your solar Ninth House. If that's impossible, take a mental journey instead via the Internet or curled up in your favorite chair with a best seller. Either one will satisfy your urge for knowledge and adventure, as will taking a short-term class for the fun of it.

Relationships
Venus enters Pisces, your solar Third House, on the 11th and aligns with lucky Jupiter midmonth. That's a plus for communication in general, and someone is likely to share great news then or at month's end. You might also make a lucky find on the sidewalk, or connect by chance with someone who can advance your career.

Money and Success
Money flows in your direction under the February 13 New Moon in Aquarius, your solar Second House. Unfortunately, so could extra expenses. Get a second or even a third quote and check references if you need a home or vehicle repair. Postpone major purchases until next month if you can.

Rewarding Days
1, 4, 6, 10, 13, 15, 16, 24, 25

Challenging Days
2, 3, 5, 9, 12, 17, 19, 23, 27

Capricorn/March

Planetary Hotspots

Difficult planetary alignments in Aries, your solar Fourth House, signal domestic or family tension. Both could occur, with one triggering the other. Your home may need a major repair that's beyond do-it-yourself; if so, get several estimates. Relocation could also be cause of family tension, or you may find it necessary to manage a relative's affairs. In addition, this is not the month to invite anyone to share your space, whether a roommate or romantic interest.

Planetary Lightspots

Mars in Leo, your solar Eighth House, resumes direct motion on the 10th. Although you won't see any immediate effect on finances, within a few weeks you can expect everything from investments to financial information to be on the upswing. Just don't expect miracles, and do continue to be cautious and conservative in money matters.

Relationships

The March 15 New Moon in Pisces, your solar Third House, continues to put a positive spin on communication. And with it comes great news, lucky news, and exciting news. The source is other people, who also spark fresh insights into life, relationships, and your career endeavors. Take the time to get acquainted with neighbors, one of whom could be your path to good fortune.

Money and Success

Your career is in focus under the March 29 Full Moon in Libra, your solar Tenth House. That's wonderful energy for a Capricorn, and you'll be totally in your element with all the initiative and incentive to stand out from the rest. You can accomplish this just by being there, but greater rewards come from snapping up every opportunity to subtly reinforce your position with decision-makers.

Rewarding Days
4, 5, 9, 10, 14, 15, 19, 26, 27, 28

Challenging Days
1, 2, 8, 11, 16, 20, 23, 25, 29

Capricorn/April

Planetary Hotspots
Stress accompanies Saturn's return to Virgo, where it aligns with Uranus in Pisces across your solar Third/Ninth House axis. A relative, in-law, or neighbor could be the source, and you should avoid legal matters if at all possible. Travel is prone to delays and cancellations, and communication in general will be strained. Drive with care and catch a ride whenever you can, especially if you're out socializing with friends.

Planetary Lightspots
With the New Moon in Aries, your solar Fourth House, on the 14th, home will be your location of choice this month. Be lazy and enjoy family evenings and weekends. Or get busy and get the group involved in spring cleaning, inside and out. Then host a get-together for friends, family, and coworkers. You might also begin thinking about summer home improvements.

Relationships
Time with friends hits the spot under the April 28 Full Moon in Scorpio, your solar Eleventh House. Earlier in the month you benefit from Venus in Taurus, your solar Fifth House. This is a great combination for social events and casual neighborhood gatherings. But you'll want to confirm dates, times, and places because Mercury turns retrograde in Taurus on the 17th. If you're wavering about a dating relationship, let things slide until mid-May when your feelings will become clear.

Money and Success
Mars in Leo energizes your solar Eighth House of joint resources, which could bring you or your mate an opportunity to earn some extra money. Be cautious, though, with investments and spending because you're more inclined now to take financial risks and also to splurge.

Rewarding Days
1, 5, 6, 8, 10, 11, 16, 20, 29

Challenging Days
2, 4, 7, 12, 15, 19, 22, 23, 25

Capricorn/May

Planetary Hotspots

You might have the urge to go somewhere this month as Jupiter in Pisces aligns with Saturn in Virgo. But this solar Third/Ninth House emphasis isn't the best planetary energy for travel, partly because reality will not live up to your expectations. Find an alternative to calm the restlessness you feel, such as a spa day, taking a class for the fun of it, or going on a nature hike with your best friend.

Planetary Lightspots

Wrap up the month with a romantic weekend rendezvous at home as the May 27 Full Moon in Sagittarius beams brightly in your solar Twelfth House of self-renewal. Togetherness time is relaxing and rejuvenating, and quality and quantity time is a rare treat. If you're single, invite a few close friends for dinner and talk the night away.

Relationships

Your social life moves into high gear after Mercury turns direct in Taurus on the 11th, followed by the New Moon in the same sign two days later. Accept every invitation, get involved in your children's activities and meet their friends' parents, and consider hosting a get-together over the holiday weekend. Some Capricorns launch a whirlwind summer romance when an acquaintance becomes much more.

Money and Success

Venus transiting Gemini, your solar Sixth House of daily work, through the 18th puts you in touch with people through your job. Some of these contacts are lucky, others try your patience. Most of all, though, you'll need to blend high energy with practical inspiration and innovation in order to win the favor of potential supporters. Go for it!

Rewarding Days

3, 7, 11, 12, 14, 15, 17, 21, 25, 30, 31

Challenging Days

2, 9, 16, 18, 22, 23, 29

Capricorn/June

Planetary Hotspots
Month's end will bring relationship tension as the June 26 Full Moon (lunar eclipse) in Capricorn clashes with several planets. Domestic or family change is possible, even likely, and you and your partner may disagree about someone moving in or out, relocation, or extensive remodeling. Again, be sure all insurance policies are up to date, and hold off if you want to ask a roommate or romantic interest to share your home.

Planetary Lightspots
Mars dashes into Virgo, your solar Ninth House, on the 7th, making this an excellent time to take a leisure-time class or to gain do-it-yourself skills. This is also your long-distance travel sector, but planetary alignments make travel unwise. Take a few days at home instead to catch up on reading and domestic projects, either of which will refresh and re-center you, body and soul.

Relationships
Your social life benefits from Mercury in Taurus, your solar Fifth House, the first nine days of June. See friends, join another family for a kids' day out, and touch base with someone faraway whom you seldom see. This person could trigger new insights that prompt you to question long-held beliefs.

Money and Success
You can expect a heavier workload as the June 12 New Moon in Gemini highlights your solar Sixth House. A financial reward is possible; praise is more likely. But you may also realize that you lack a skill necessary for further advancement. Make the effort to get the knowledge you need, whether on your own, through your company, or online resources.

Rewarding Days
4, 5, 8, 9, 13, 14, 17, 18, 22, 27

Challenging Days
3, 6, 10, 11, 19, 21, 24, 25, 26

Capricorn/July

Planetary Hotspots

Saturn spends its final days in Virgo before returning to Libra, your solar Tenth House, on the 21st, where it aligns with Uranus in Aries, your solar fourth House, on the 26th. This lineup signals career or domestic change, or possibly both. Downsizing is possible for you or your mate, as is relocation. Again this month, it would be unwise to welcome a new roommate, and you should delay a home purchase if that's in your plans.

Planetary Lightspots

The last two weeks of July favor travel if you want to get away for a few days. With Venus in Virgo, your solar Ninth House, from the 10th on, the timing is great for a romantic holiday or a family weekend trip. Opt for an inexpensive destination, and consider one that's close to home. A change of scenery is what you need.

Relationships

Relationships get a boost from the July 11 New Moon (solar eclipse) in Cancer, your solar Seventh House. Close ties deepen, and you'll also have the opportunity to add new faces to your circle, possibly while out of town. Couples are on the same wavelength as Mercury transits Cancer through the 8th, and the same will be true of relatives. Some singles launch a new romance with someone from the past.

Money and Success

The July 25 Full Moon in Aquarius, your solar Second House, accents finances, including earnings and spending. The lunar energy connects with Saturn in Libra and Jupiter and Uranus in Aries. That could benefit your bank account, but extra expenses are also possible. Forgo major purchases until year's end, if possible.

Rewarding Days

1, 2, 6, 7, 11, 14, 15, 17, 24

Challenging Days

3, 9, 10, 16, 19, 20, 22, 23, 30

Capricorn/August

Planetary Hotspots
Events of the past few months culminate now as four planets in your solar First, Fourth, and Tenth Houses are activated. This alignment could trigger career and domestic changes again this month, including downsizing and relocation. You'll also encounter difficult, controlling people. What you'll find especially challenging with this planetary influence, however, is that you'll experience global conditions on a personal level and feel as though you're swimming upstream. Take charge of what you can, and take a fresh look at your life as a whole in the context of positive transformation.

Planetary Lightspots
Your curiosity comes alive under the August 24 Full Moon in Pisces, your solar Third House, which also puts you in the communication loop. With Uranus returning to Pisces on the 13th, your mind will be quick and learning easy. Flashes of insight are also associated with Uranus, so listen to your sixth sense.

Relationships
Mix-ups and misunderstandings will be more the norm than the exception after Mercury turns retrograde in Virgo, your solar Ninth House, on the 20th. If you can, postpone travel because it'll be especially prone to delays and cancellations. If not, plan ahead as best you can. And don't be surprised if your thoughts turn inward as you search for answers. Meditation can help you access your true feelings.

Money and Success
This month's financial emphasis is Leo, your solar Eighth House of joint resources, where the New Moon occurs on the 9th. This positive influence can be an asset for your partner's income as well as communication with lenders. However, be sure to read all the fine print before you sign any document.

Rewarding Days
2, 3, 7, 11, 15, 21, 23, 26, 29, 30

Challenging Days
4, 6, 10, 12, 13, 19, 20, 25, 27, 28

Capricorn/September

Planetary Hotspots
Although this summer's planetary influences begin to wane, Pluto in Capricorn will be activated the first week of the month and again around the time of the Full Moon in Aries, your solar Fourth House, on the 23rd. Family and domestic issues will arise, and there could be difficulties with relatives. Pause and think calm thoughts before you speak and act. This is a test of your will as well as your ability to adapt to a changing situation. And although specific events may be out of your control, you do have the determination to stand firm now.

Planetary Lightspots
Jupiter returns to Pisces on the 8th and merges with Uranus in the same sign on the 18th. This lineup of good fortune could trigger a lucky moment, a chance encounter with someone who could open a door of opportunity. You could meet this person near home, so get acquainted with neighbors and also be alert while running errands.

Relationships
Venus enters Scorpio, your solar Eleventh House, on the 8th, followed by energetic Mars on the 14th. This planetary duo is great for your social life and time with friends. A vacation trip or weekend getaway with pals around the New Moon in Virgo, your solar Ninth House, on the 18th could also be your link to a lucky contact.

Money and Success
Your solar Tenth House is an active sector this month as several planets advance in Libra. Career matters will be frustrating, uplifting, challenging, and rewarding, depending upon the day, and month's end could bring positive developments. But resist the urge to push yourself and others. A go-with-the-flow attitude will get you further.

Rewarding Days
2, 6, 7, 8, 10, 12, 17, 20, 21, 22, 27

Challenging Days
1, 3, 4, 9, 11, 15, 16, 23, 25, 29, 30

Capricorn/October

Planetary Hotspots
Life is far more calm in October than it has been in recent months, and most everything will go according to plan. There is an exception, however. Venus turns retrograde in Scorpio, your solar Eleventh House, on the 8th. This six-week influence can slow your social life and prompt you to decline meet-and-greet invitations in favor of cozy evenings at home. It may also have an effect on workplace communication, and although you'll want to work solo, remind yourself to keep others in the loop.

Planetary Lightspots
Home will definitely be your location of choice under the October 22 Full Moon in Aries, your solar Fourth House. This positive influence is as great for family gatherings as it is for completing all those unfinished domestic projects and repairs. Also devote some time to cleaning out closets and more; it will bring you great satisfaction.

Relationships
You'll still have opportunities to socialize despite the influence of retrograde Venus. But expect events to be more low-key and with close friends, which will suit you fine now. Consider hosting a small group around the Full Moon, and also set aside time for a lazy evening with your best pal to talk the night away. Share your hopes and wishes and listen closely to the feedback.

Money and Success
Again this month your solar Tenth House is in high focus, especially the week of the October 7 New Moon in Libra, your career sign. You could earn a raise or bonus, or be asked to take on additional responsibilities. If you're job-hunting, the New Moon is great for sending out résumés.

Rewarding Days
1, 5, 7, 10, 15, 17, 18, 19, 24, 26, 29

Challenging Days
2, 4, 6, 9, 13, 14, 20, 21, 27, 28

Capricorn/November

Planetary Hotspots
Life continues along the same mostly easy path this month. But with Mars in Sagittarius, your solar Twelfth House, you'll want to be sure to get plenty of rest and sleep. Otherwise, the planetary influences could trigger a cold or flu, which would definitely put a damper on what promises to be an upbeat month as you head into the holiday season.

Planetary Lightspots
Although it will be month's end before this influence takes hold, you'll have a way with words that extends into December, thanks to Mercury in your sign. That's a plus for communication in every area of life, and you also should give some thought to personal plans and goals for 2011. Then you'll be ready to act on them once January arrives.

Relationships
November's New Moon in Scorpio on the 6th and Full Moon in Taurus on the 21st are in sync with your social life. With both these sectors active this month, you'll have to plan ahead and closely track your schedule in order to keep up with it all. Friendship gatherings are among the best, and some singles will launch a whirlwind romance. You might also consider hosting a group of friends for a laid-back Thanksgiving at your place.

Money and Success
Career benefits could come your way once Venus returns to Libra, your solar Tenth House, on the 7th and then resumes direct motion on the 18th. This fortunate influence could set the stage for gains next month, so you should take advantage of every opportunity to cultivate supporters and showcase your skills and talents.

Rewarding Days
1, 2, 5, 6, 12, 14, 16, 19, 20, 28

Challenging Days
3, 7, 9, 10, 11, 17, 23, 24, 27, 30

Capricorn/December

Planetary Hotspots

Frustration can get the best of you this month when personal plans are put on hold. That's one effect of Mercury, which turns retrograde in Capricorn on the 10th. Your best bet is to go with the flow and forget it rather than trying to push things forward. Retrograde Mercury also cautions against purchasing electronics and other big-ticket items. If it's a must, consider an extended warranty.

Planetary Lightspots

Mars lends you its high energy from the 7th on, when it will be in your sign. That's great for a busy lifestyle and keeping up the pace, but slow down enough to prevent distractions which can lead to mishaps. This is especially important—and easier to achieve—around the time of the December 5 New Moon in Sagittarius, your solar Twelfth House of self-renewal.

Relationships

Just in time for the holidays, Venus spends the entire month in Scorpio, your solar Eleventh House of friendship. Line up dates, outings, and get-togethers with pals, and snap up every invitation that comes your way. This is a very positive networking influence that can put you together with the right people at the right time.

Money and Success

Your job will keep you on the go at month's end when the December 21 Full Moon (lunar eclipse) in Gemini spotlights your solar Sixth House. Aim to complete your to-do list by year's end so you can get a fresh start in January, and also include a few coworkers in your holiday social plans. Be sure, though, to check and double-check all work output for errors. You'll find them more easily when Mercury is retrograde and that can prevent an embarrassing moment next month.

Rewarding Days

1, 3, 4, 8, 10, 12, 13, 17, 26, 29, 30, 31

Challenging Days

6, 7, 11, 14, 15, 20, 21, 24, 27, 28

Capricorn Action Table

These dates reflect the best—but not the only—times for success and ease in these activities, according to your Sun sign.

	JAN	FEB	MAR	APR	MAY	JUN	JUL	AUG	SEP	OCT	NOV	DEC
Move			18–31									
Start a class		15	2–19		7							
Join a club			4, 5						11, 12			2, 3
Ask for a raise	25	3	11		5							
Look for work	25, 26		21, 22		14	10–20						
Get pro advice		24				25–30	1–8			1		
Get a loan			25, 26			15	12, 13					
See a doctor						11						
Start a diet	25											
End relationship								7	3, 4, 30	1	24	
Buy clothes			19	5, 6, 15–19	12–14	9		3				
Get a makeover			9, 10	5					16, 17			
New romance					12, 13	9		2, 3				
Vacation							28–31	1–6	16–22			

AQUARIUS

The Water Bearer
January 20 to February 19

Element: Air
Quality: Fixed
Polarity: Yang/masculine
Planetary Ruler: Uranus
Meditation: I am a wellspring of creativity
Gemstone: Amethyst
Power Stones: Aquamarine, black pearl, chrysocolla
Key Phrase: I know
Glyph: Currents of energy

Anatomy: Ankles, circulatory system
Color: Iridescent blues, violet
Animal: Exotic birds
Myths/Legends: Ninhursag, John the Baptist, Deucalion
House: Eleventh
Opposite Sign: Leo
Flower: Orchid
Key Word: Unconventional

Your Strengths and Challenges

Your free-spirited, independent nature is unmistakable. But not as obvious, and what periodically emerges, is your more conservative side. This contributes to your reputation for the unpredictable. As soon as people think they have you figured out, you switch gears in tune with Uranus, your ruling planet. You can be determined as well as stubborn, depending upon the situation, and change is something you shy away from—unless of course it's your idea. Initiating change is one of your strengths, but it can be detrimental if your motive is simply to upset the status quo to keep life interesting. So use your abilities where they'll have the greatest positive impact.

Aquarius is the sign of the humanitarian, and you're a champion of individual rights and equality. You welcome people from all walks of life as long as they pique your interest, but you become emotionally involved with only a few. More comfortable in the mental realm, you're sometimes seen as distant and aloof, and anger can trigger an icy gaze. Your intuitive, inventive mind often senses approaching trends, sometimes years in advance, and some Aquarians have a unique mechanical ability that helps them create new products. Intuition also helps you keep your life on track into the future, with flashes of insight that signal important turning points.

Your Relationships

Born under the universal sign of friendship, these people have a special place in your life. Your friends, whether male or female, come from diverse backgrounds and status levels and you accept each as an individual, based solely on his or her qualities. Everyone in your circle shares a single trait: each is unique in some way, interesting and outside the norm. You find regular people boring, and only those who fascinate and intrigue you become your friends.

You delight in a lively romantic life, with plenty of flirting and spur-of-the-moment activities. Playing the field and even simultaneously dating two people keep things interesting, and long talks are one way to your heart. Because independence is all-important to you, this trait could be your stumbling block to commitment. But with the right person, someone who's also your best friend, you're willing to become a partner—as long as he or she understands your need for freedom. You could make a sensational match with a Leo, your opposite sign, and Aries and Sagittarius are in sync with your free-spirited outlook. You would have much in common with another Aquarius, or the other air signs, Gemini and Libra, but Taurus and Scorpio could be too possessive for you.

Your family is your comfort zone, and in this area of life, you're surprisingly traditional. Given a choice, you'll stay close to home in adulthood, where your children can form close ties with your extended family. Aquarians usually have small families, and twins are possible. As a parent you nurture your children's curiosity and encourage learning. But you also can be inconsistent in your message, so try to remember to reinforce core principles that will establish a strong foundation for their adult lives.

Your Career and Money

You're attracted to a mentally challenging career field, and it's also a must that you're in a field with leadership opportunities and one where you can initiate change on some level. A calm workplace environment is essential for high productivity, and you're a loyal employee who operates best in a team setting that emulates family. But that can make it tough to cut ties and move when you've outgrown your position. Remind yourself that you'll form the same strong relationships wherever you go.

Your financial success is directly linked to planning, organization, and budgeting. Otherwise, money can slip through your fingers. But you also usually attract what you need when you need it. Home ownership can be very profitable, as can long-term investments if you take a conservative approach and pay attention to details. You also have a soft spot for people in need. That's great, but put you and yours first, and help others help themselves by directing them to other resources.

Your Lighter Side

You have a passion for gadgets of all kinds, especially electronic ones. Their usefulness, while important, isn't as much of a factor as just having them because they appeal to your innovative mind. In fact, you could invent the next must-have gadget or use the ones you own to create an easier, simpler way to do a routine task.

Affirmation for the Year

Knowledge is my guiding light.

The Year Ahead for Aquarius

January 17 puts money in the spotlight as bountiful Jupiter enters Pisces, your solar Second House of personal resources. This transit has all the potential to boost your bank balance in 2010, so think positive and attract abundance. Jupiter here does come with a challenge, however: spending. It can rise as fast (or faster) than income if you're not careful. Save and pay off any accumulated debt first and aim for more in the bank at year's end than you had at the beginning. Another good reason to save is to cover any unexpected expenses because Jupiter can multiply those as well. This is an excellent year to become a better informed consumer and to educate yourself on financial planning and money management. Jupiter in the Second House also encourages you to look beyond the materialistic realm. Define what you value, what is ultimately important to you. Then use that knowledge to devise a plan to achieve what you want from life.

Jupiter makes a brief visit to Aries, your solar Third House, June 5–August 12, before returning to Pisces for the rest of the year. Expect a fast-paced summer with this transit because you'll be on the run, juggling an unusually hectic schedule. So much so that it will be a challenge to stay organized. Expect increased contact with relatives and neighbors, and a heavy volume of calls and e-mail. Suddenly, everyone will want their share of your time and attention—and they'll all want it now! Be prepared to set limits, and promise yourself thirty relaxing minutes a day to read, walk, or meditate.

Saturn also spends time in two signs this year, briefly returning to Virgo, your solar Eighth House of joint resources, April 7–July 20 to complete its transit of that sign. This is your opportunity to finish any unfinished business begun as long ago as 2007, when Saturn entered Virgo. Like Jupiter in Pisces, the emphasis here is on money, but from the opposite perspective: lenders, your partner's income and debt, insurance, inheritance, and long-term investments. If debt has been (or is) an issue, now is the time to resolve these matters by making final payments or setting plans in motion to reach that goal. Be proactive and responsible, which is Saturn's main message. Anything less will require another attempt to learn these financial lessons seven or fourteen years from now.

A long-awaited or contested insurance settlement or inheritance could be finalized now. In either case, you're unlikely to get more if you continue to negotiate. Know when to quit and move on. It's also possible you could see reduced health or other insurance benefits, or there may be a large premium increase for the same coverage. This Saturn transit could also reflect a downward trend in your partner's income or long-term investments, and you should avoid taking on new debt if at all possible.

Saturn begins the year in Libra, your solar Ninth House, and returns there after its final Virgo transit. You're more in sync with this Saturn transit because it occurs in an air sign. There will be challenges; they'll just be easier to manage. Placed in the Ninth House, Saturn emphasizes knowledge and education. So you should seriously consider starting or finishing a degree or studying for a higher level of certification within your career field. True, it will require commitment and restrict free time, but it will be well worth it when Saturn enters your solar Tenth House of career. What you do now will reap rewards then. This is also the time to assess your career direction regarding future and long-term employment. Check out the options if you're at all interested in switching careers.

Difficulties with in-laws are possible, or you might need to manage a relative's affairs. Business travel will keep some Aquarians away from home for extended periods, and the same could be true if a family member needs your assistance.

Knowledge, however, goes beyond book-learning when Saturn is in the Ninth House. Open your mind to other cultures and opinions and truly listen to what people say. This is easier for you than for some Sun signs, but even you can have fixed ideas. So look beyond what you've always considered the norm, ask questions, and explore your beliefs and assumptions. Some will weather the test, and others will disappear into the past. On a deeper level you can use this transit to explore your psyche, possibly with the help of a counselor or life coach.

Uranus also transits two signs this year, spending most of 2010 in Pisces, your solar Second House. No doubt your finances have been up and down during the years since Uranus entered this sign. That's the nature of this quirky planet. This year, just as it did in 2008 and

2009, Uranus forms an exact alignment with Saturn in Virgo across your solar Second/Eighth House axis. Stresses and strains involving money matters are likely, which is all the more reason to be thrifty and practical as well as proactive if you see financial difficulties on the horizon. Take care of you and yours.

The upside of this transit is that you could luck into a nice windfall or have one of those brilliant Aquarian moneymaking ideas. Pursue your idea, but don't tap your credit to fund a business proposition, no matter how great you feel about it. Look instead to a friend or relative who's in a position to do that for you.

Uranus briefly visits Aries, your solar Third House, May 27–August 12. This transit adds the element of surprise and lucky contacts are likely and even probable in June, when these two planets merge their energies. And if you're awaiting good news, it's likely to arrive then. Beware, though, of people who promise you the world; they're unlikely to deliver. Uranus in the Third House also urges caution on the road because accidents can happen in a flash through no fault of your own. Postpone a vehicle purchase unless it's an absolute necessity. You'll also want to be diligent about backing up computer files, and to be extra careful with anything electrical.

This Uranus transit is excellent for insights and intuition. Pay attention to both because they're likely to plant seeds for the future which will bloom in 2011. You'll be an especially quick study this year, so consider doing something that will satisfy your urge for learning. Try something new, something that's always been of interest. Who knows? It could set you on a whole new career path. At the least you should read and explore Internet resources as an outlet for your curiosity, which will come alive this year.

Neptune continues its long Aquarius transit, having entered this sign in 1995. Tap into the inspirational nature of this planet and let it motivate you to strive for and achieve personal goals. But be sure to give yourself a periodic reality check because Neptune is also the planet of illusion. Even if you feel you're on target it's wise to bounce ideas off someone you trust who's also practical. One of the best uses of this transit is to help others. But, again, you'll need to keep things in perspective because there's a fine line between being supportive and being an enabler. Along the same lines, you're more than usually susceptible to sob stories now, so adopt a policy of hesitation. Think things through before you agree to a request, especially one that involves money. The best choice, given this year's other planetary influences, is to help others find resources rather than to give away your own.

Pluto, the slowest moving planet in the zodiac, will be in Capricorn, your solar Twelfth House, through 2024. Its influence in your self-renewal sector will be subtle yet profound as it works through your subconscious, but more prominent during the year it contacts your Sun. Memories, regrets, and unresolved issues will emerge for one purpose: so that you can come to terms with them. Some will be positive; others will be negative. All, however, will have some meaning in the context of your current life. And now you'll be better prepared to deal with them, having gained life experience in the intervening years.

This transit will push you to examine your psychological motivations, including what holds you back. Think about how you react in stressful situations, especially those involving other people, and decipher the reasons behind your responses. By doing this you will learn much about yourself and thus become more effective in the outer world and in your career.

What This Year's Eclipses Mean for You

This year features four eclipses, two solar and two lunar, each of which is in effect for approximately six months. The first two occur in Capricorn, your solar Twelfth House, reinforcing your current need for time alone to delve into the depths of your mind. Eclipse energy is active for approximately six months.

The January 15 solar eclipse in Capricorn is the more favorable of the two, aligned with Venus, also in Capricorn, and Uranus in Pisces, your solar Second House. Cross your fingers. You could profit from a lucky find while walking down the street or spot treasure at a yard sale or thrift store. It's even possible something from your childhood could be a valuable collectible. You can also put this eclipse to good use cleaning out closets; take the best to a consignment shop for a little extra cash.

The June 26 lunar eclipse in Capricorn will activate Pluto in Capricorn and Mercury in Cancer, spanning your solar Sixth/Twelfth House axis. This indicates tension and change in the workplace as well as the potential for power plays. Do your best to steer clear of them while also protecting your job. You'll also have an opportunity to create and promote innovative ideas and fresh approaches; just be sure to line up support before you go public with them. Equally possible is an opportunity to expand your influence, but this will require finesse so you don't step on toes.

Balancing the June eclipse is the July 11 solar eclipse in Cancer, which could trigger a moneymaking opportunity, particularly if you're willing to take on extra work. You can use this eclipse in another way as well: to work around the house. Get the do-it-yourself skills you need for minor repairs or improvements. It will be time well invested whether you hope to sell your home or plan to stay. Also take the time to organize important papers and financial records.

The energy shifts to Gemini, your solar Fifth House, with the December 21 lunar eclipse. This eclipse focuses your attention on your children, and some Aquarians will welcome an addition to the family. Friends are also highlighted by this eclipse, and you can widen your circle by getting acquainted with your children's pals and their parents. Creativity and romance are also in this eclipse's spotlight, and a chance encounter will be the start of a new relationship for some Aquarians.

Saturn

If you were born between February 16 and 18, Saturn will contact your Sun from Virgo, your solar Eighth House of joint resources, between April 7 and July 20. You, among all Aquarians, will need to take a serious look at finances. But rather than worrying, which is to be expected with this connection, take action and deal with any challenges. Professional advice can be beneficial now, whether to handle debt or to know how best to manage retirement funds and investments. Be realistic if you experience a change in family income and develop a budget to maximize available funds. The answer is to be proactive in all money matters even if you'd rather avoid them.

This Saturn/Sun connection can also manifest as concern about the future even though your job and family income are status quo. Put yourself on a strict budget, pay off any accumulated debt, and save for unexpected expenses and retirement. The tension will begin to ease as you see your bank balance rise.

Stress and strain are also likely if you're awaiting on an insurance settlement or inheritance (or settling an estate), either of which could be less than you hoped for. That's all the more reason to avoid spending until you have the cash in hand.

If you were born between January 20 and February 5, Saturn in Libra, your solar Ninth House, will contact your Sun this year. But **if your birthday is after January 24**, you'll experience this Saturn transit for only a few weeks between mid-August and year's end. For Aquarians born before then, Saturn will be active in your life between January 1 and April 6, and again from July 20 through early August.

You'll be able to tap the best of Saturn with its easy connection to your Sun, using all the practicality of this serious planet to fuel your determination and ambition. Transiting the Ninth House, Saturn encourages you to expand your knowledge base whether through self-understanding, casual learning, or by pursuing a degree. What is most important here is to grow both as an individual and as someone who is in touch with the world at large by exploring your life philosophy and how you fit into the bigger scheme of things.

Appropriately, religion and spirituality are also associated with this sector. You may return to or reject what you learned as a child, or embrace an entirely new set of beliefs to guide your daily life. Part of this is, of course, a new maturity (whatever your age) based on an understanding of the past to prepare yourself for the future. And although this can be an easy process overall, it's also bound to be challenging because Saturn will contact Pluto in Capricorn. So, the more deeply you delve into your subconscious, the more all-encompassing this process and the results will be.

Uranus

If you were born between February 12 and 19, Uranus will contact your Sun from Pisces, your solar Second House of personal resources—your income, possessions, and spending habits. Although change is a given wherever Uranus is involved, this transit won't necessarily affect your income. But it could, especially if Saturn will also contact your Sun this year. The wise choice is thus to conserve resources—be thrifty, avoid credit as much as possible, and save a higher percentage of your income. Take a close look at how you spend and on what, and then find ways to cut back where necessary and to get the most for your money. Do the same with savings, and have a set amount automatically deposited in an account reserved for emergencies. All this is more than doable this year because money will flow your way, thanks to Jupiter in Pisces.

This Uranus transit is one of the best for ridding yourself of unneeded or unwanted items, and you probably can net a surprisingly large profit from your discards. But that's the practical, outer level of the changes occurring within as you sort through what you value personally, materialistically, and in your life as a whole. Define this before you toss out everything you own—or spend every cent you have!

If you were born January 20 or 21, you're among the first of your sign to have Uranus contact your Sun from Aries, your solar Third House (May 26–August 12). Your mind will work overtime, with rapid-fire ideas. You'll grasp concepts in a flash and do the same with decisions. That can be a real plus in certain situations, but try to remember that snap decisions can lead you astray. Retreat and reflect before making major moves; otherwise you may regret.

All this is because you'll have a strong urge for change. You can achieve the same goal (or nearly so) through learning, either in a formal setting or on your own, or by being a positive change agent in your community. Stir the pot! For example, you could organize a citizens or neighborhood group to refurbish a playground or park, or spearhead a campaign to expand your local library. In a larger sense, what you do to benefit those around you will help you redefine yourself, which is what this Uranus-Sun contact encourages you to do.

Neptune

If you were born between February 12 and 19, you'll have an extra special aura this year, thanks to Neptune, which merges its energy with your Aquarius Sun. This magical combination is terrific if you're in search of a new romance. Just don't take things too seriously or to the commitment level because what feels like true love may be only an illusion—being in love with love.

In addition to boosting your charm and powers of attraction, Neptune will trigger your sixth sense, prompting your inner voice to speak loudly. The more you listen the greater its strength, so nurture this side of yourself. Also take note of dreams, which can be insightful and even prophetic. Neptune is also the ultimate planet of creativity. Find an outlet to express this side of yourself, which will also increase your intuition.

But Neptune's influence can have you feeling as though you're in a fog, drifting and directionless, with ideas and plans shifting like the wind. So major life decisions aren't wise this year, when you'll also be idealistic, hopeful, and trusting. Skepticism can be your greatest ally when dealing with people and situations. Ask questions and accept only factual, concrete answers.

Pluto

If you were born between January 22 and 25, Pluto in Capricorn, your solar Twelfth House, will contact your Sun this year. Transiting your self-renewal sector, Pluto encourages you to assess your life and lifestyle—what works and what doesn't, what you want to change and what you want to retain. Be honest with yourself. Are you as healthy as you could be? With Pluto in place, you'll have the willpower to change almost anything, even long-standing habits. Pluto here also operates on a mental level, targeting your subconscious. So don't be surprised if memories and long-dormant desires emerge into your thoughts. Explore each one, resolve regrets, and take action to pursue the dreams that recapture your interest. Let intuition guide you here.

Jupiter-Saturn-Uranus-Pluto

If you were born between January 20 and 23, this four-planet alignment will contact your Sun this summer. All bets are off when Uranus is involved. You can expect events to quickly occur and encompass your solar Twelfth House (Pluto in Capricorn), Third House (Jupiter and Uranus in Aries), and Ninth House (Saturn in Libra). Sudden travel concerning a relative, or perhaps for an extended period, is possibly. Extensive business travel is also possible. A community or neighborhood activity could thrust you into the limelight, and there could be difficulties with a neighbor or homeowners association. Legal action is also possible with these planets, and you'll want to be sure to drive with extra care and to always socialize with a designated driver.

A chance encounter could trigger an exciting opportunity, but be aware that initial enthusiasm may quickly wane or be put on hold until next year. So don't make major decisions based on promises or guarantees of success.

Aquarius/January

Planetary Hotspots

Someone at a distance could make your life difficult this month, possibly a friend, possibly a relative or in-law. That's the best reason to screen calls and postpone a visit if you can. But with this month's New Moon (solar eclipse) in Capricorn, your solar Twelfth House, on the 15th, it's a great time to resolve lingering issues. Browse the self-help section and use your intuition. Dreams can be insightful.

Planetary Lightspots

Circle the 18th on your calendar. That's the date Venus enters your sign, bringing with it an extra level of magnetic sparkle to charm anyone you meet. Your powers of attraction also get a boost. Make the most of it by truly believing you deserve exactly what you want. Then make it happen!

Relationships

Relationships will be a bit bumpy at times this month (and through mid-March) as Mars travels retrograde in Leo, your solar Seventh House. You'll especially want to be prepared to curb your temper and your impatience at month's end. Try to see things from the other person's perspective and keep an open mind. At the least, agree to disagree and find a compromise that works for both of you.

Money and Success

Money! It starts flowing your way on the 17th, when Jupiter enters Pisces, your solar Second House of personal resources. The year-long trend can do wonders for your bank balance, but only if you stick to a budget and seldom give in to the strong temptation to spend. Make saving a priority because unexpected expenses are possible, even likely, as the year unfolds.

Rewarding Days
1, 2, 3, 7, 11, 12, 15, 17, 22, 25

Challenging Days
5, 6, 8, 13, 16, 20, 23, 27, 31

Aquarius/February

Planetary Hotspots
With several planets in your sign, life is on your side this month. That's a great reason to celebrate. Retrograde Mars in Leo, however, is not. Last month's relationship pattern continues, although tension will be less pronounced. Instead, you'll have difficulty connecting with people, and some will resist your attempts to promote partnership efforts. That's to be expected, so give people the space they need rather than trying to push them.

Planetary Lightspots
The February 13 New Moon in Aquarius is all about you and your hopes, wishes, desires, and goals. Put Mercury in your sign from the 10th on to good use and contemplate what you want to achieve in the next twelve months. With concrete goals and a solid plan, you can achieve almost anything. Go for it!

Relationships
Although Mars signals relationship ups and downs this month, you'll also be in sync with many. That's especially true after Mercury enters your sign on the 10th. Don't hesitate. Strike up a conversation with anyone who catches your interest. That person could be your lucky charm, the one who sparks an idea or new insights into a puzzling personal dilemma. Just don't blindly accept what you hear. Weigh the pros and cons, and make your own decision.

Money and Success
You'll have an opportunity to increase earnings the week of the February 18 Full Moon in Virgo, your solar Eighth House. The extra cash could come through your job, friend, or as a result of wise investments. But you'll want to postpone major financial decisions because it will be tough to see the downside as well as the upside. Optimistic enthusiasm could get the best of you.

Rewarding Days
3, 4, 7, 8, 11, 13, 14, 18, 26

Challenging Days
2, 5, 6, 9, 12, 17, 19, 20, 23

Aquarius/March

Planetary Hotspots
Communication is this month's hotspot. Difficult planetary alignments in Aries, your solar Third House, can trigger conflict with relatives or neighbors, so do your best to avoid these people. If necessary, end the call or walk away. Also be very careful behind the wheel, don't speed even if you're late, and catch a ride when you're out socializing. Take time out when pressure builds, and look within to identify the underlying issue that upsets and aggravates you.

Planetary Lightspots
Take advantage of the March 29 Full Moon in Libra, your solar Ninth House, when it encourages you to plan a brief escape. Travel isn't the best idea because of potentially challenging planetary alignments, but you can achieve the same end at home. Take a quick online class, look for information on the Internet, or curl up in your favorite chair with a best seller.

Relationships
With a few exceptions, relationships begin to get back on track after Mars in Leo, your solar Seventh House, resumes direct motion on the 10th. Open your mind to alternative viewpoints and take the initiative to talk out recent misunderstandings. Aim for compromise; if that's not possible, at least agree to disagree with a smile.

Money and Success
Cross your fingers! Beneficial planetary alignments and the March 15 New Moon in Pisces, your solar Second House, enhance the possibility of a lucky win, raise, bonus, or unexpected windfall. Treat yourself to something you covet, and stash the rest in savings. That way you won't be tempted to splurge.

Rewarding Days
4, 5, 9, 10, 12, 13, 14, 15, 17, 26, 30

Challenging Days
2, 3, 8, 11, 16, 18, 20, 23, 25, 29

Aquarius/April

Planetary Hotspots
Finances aren't nearly so positive this month as Saturn returns to Virgo and aligns with Uranus in Pisces across your solar Second/Eighth House axis. The duo can trigger increased expenses, a decrease in income or investments, and mix-ups with credit and loans. Check your (and the family's) credit reports and handle any errors. Do the same with statements. Try to postpone major purchases, but if it's a necessity, consider getting an extended warranty.

Planetary Lightspots
The April 14 New Moon in Aries, your solar Third House, encourages you to dash out of town for a long weekend. Visit nearby friends, head for a relaxing destination, or check into a local luxury hotel. Where you go isn't as important as getting away and taking a break from daily life. You could even do the same at home; turn off the phone and get lost in a favorite hobby.

Relationships
Family relationships are generally positive. But keep track of schedules and carefully choose your words because Mercury turns retrograde in Taurus, your solar Fourth House, on the 17th. You'll also want to do a quick check of your home every few days just to be sure appliances are running as they should.

Money and Success
You're in the career spotlight of the April 28 Full Moon in Scorpio, your solar Tenth House. Aim to achieve and showcase your skills and talents at every opportunity. Also be sure to double-check projects and other work before you call it finished. The only challenge here is finding a way to manage career and family responsibilities. Strive for a balanced lifestyle.

Rewarding Days
2, 3, 8, 13, 14, 17, 21, 26, 27, 30

Challenging Days
4, 7, 12, 15, 19, 22, 23, 25, 28

Aquarius/May

Planetary Hotspots
To spend or not to spend will be one of your financial dilemmas this month as Jupiter in Pisces aligns with Saturn in Virgo, your financial sectors. Ask yourself if you really need whatever it is. If the answer is yes, wait a few days to see if you change your mind; you probably will. On another level this alignment urges you to take a serious look at money matters, including your financial goals, budgeting, spending habits, and income potential. Devise a plan to change what needs changing and then implement it.

Planetary Lightspots
You're drawn to the domestic scene under the May 13 New Moon in Taurus, your solar Fourth House. Family relationships get back on track after Mercury turns direct in Taurus on the 11th. This is a great month for household projects, repairs and improvements. Go for do-it-yourself and do it on a budget. You'll like the results and the satisfaction of a job well done.

Relationships
The May 27 Full Moon in Sagittarius, your solar Eleventh House, promises to enhance your social life the following two weeks. Friendship takes center stage and a lucky moment could connect you with a new romantic interest or someone who can advance your career. The lunar energy is also a plus for group activities; consider getting involved in a good cause.

Money and Success
You're popularity-plus on the job as Venus transits Cancer, your solar Sixth House, from the 19th on. Don't take it too far, though, because even benefic Venus can do only so much. Match this fortunate influence with high productivity and leave socializing (and romance) with coworkers to after hours and weekends.

Rewarding Days
1, 5, 6, 10, 11, 14, 15, 19, 24, 27, 28

Challenging Days
2, 9, 13, 16, 18, 22, 23, 29

Aquarius/June

Planetary Hotspots
The June 26 Full Moon (lunar eclipse) in Capricorn, your solar Twelfth House, triggers stress and strain in the workplace because the lunar energy also activates Pluto in Capricorn and several other planets. Expect power plays and controlling people, which you should do your best to avoid. Getting drawn into these conflicts could result in a setback, so it's important to choose your battles with care. Be wary. Someone who appears to be a friend could be the opposite.

Planetary Lightspots
Jupiter enters Aries on the 6th, where it joins Uranus in Aries on the 8th. This marks what could be a sensational week featuring a chance encounter. Someone, possibly a neighbor or relative, might be your lucky charm for a career connection. You also could feel the instant rapport of a friendship, a connection that's meant to be, that in some way changes your life.

Relationships
June brings the New Moon in Gemini, your solar Fifth House, on the 12th, just days after Mercury's arrival there. This influence, combined with Venus in Leo, your solar Seventh House, has relationships in positive territory. You'll be in sync with most everyone, and singles could make a match that eventually results in lasting commitment. The lunar energy also highlights your children, so you'll want to spend extra time with them this month.

Money and Success
Mars transiting Virgo, your solar Eighth House, from the 7th on, gives you extra incentive and initiative to increase your earning power. But it also can trigger extra expenses and the urge to splurge. Keep an eye on your budget and set shopping limits.

Rewarding Days
1, 7, 11, 15, 16, 20, 27, 28, 29, 30

Challenging Days
2, 3, 5, 6, 10, 12, 19, 21, 24, 25

Aquarius/July

Planetary Hotspots
Tension builds as Saturn returns to Libra on the 21st and aligns with Uranus in Aries on the 26th. This emphasis on your solar Third/Ninth House axis advises caution on the road and with difficult people, especially neighbors. Legal action may have an unexpected outcome. It's also possible you could be selected for jury duty in an extended, high-profile case. Reschedule travel if you can as it will be prone to delays and cancellations.

Planetary Lightspots
You're at the center of the Universe as the July 25 Full Moon shines brightly in your sign. Make the most of your attention-getting power and delight in every moment. But also take time to review how far you've come since January and what you want to achieve the rest of the year. Then enlist the support and help of others to accomplish your goals.

Relationships
Relationships are also featured this month, thanks to Venus in Leo, your solar Seventh House, and Mercury in the same sign from the 9th to the 26th. People will be especially receptive to you and your requests now, and you can easily establish rapport with almost everyone. Someone you meet around the 10th or 22nd could be your lucky charm. Introduce yourself to neighbors.

Money and Success
You could earn a raise when the July 11 Cancer New Moon (solar eclipse) accents your solar Sixth House. But the lunar energy could also trigger workplace changes that give you extra responsibilities. Consider this an opportunity, and listen closely if a supervisor suggests you further your education. It will pay off well in a few years.

Rewarding Days
4, 5, 8, 12, 13, 17, 18, 26, 27, 31

Challenging Days
2, 3, 9, 10, 16, 21, 22, 23, 30

Aquarius/August

Planetary Hotspots
Last month's events reach a peak now as Jupiter, Saturn, Uranus, and Pluto interact in your solar Third, Ninth, and Twelfth Houses. Expect the tension to build in early August and peak midmonth. Continue to be cautious on the road and in legal matters, and do your best to avoid travel and difficult people. Overall, your wisest choice this month is to maintain a low profile as much as possible.

Planetary Lightspots
Despite this month's stresses and strains, you'll also have many upbeat moments as planetary alignments put you in the forefront of activity. Seek information and knowledge, both of which can change your view of the world and your place in it. And don't be surprised if you suddenly gain new insights into what motivates you and what holds you back.

Relationships
You'll have an opportunity to connect with many people under this month's New Moon in Leo, your solar Seventh House, on the 9th. Some will inspire you, but others may tell you only what you want to hear. Use your intuition, ask questions, and expect answers. Be cautious if you need to consult a professional. Check credentials.

Money and Success
The August 24 Full Moon in Pisces, your solar Second House, is promising for money matters. But because Mercury turns retrograde in Virgo, your solar Eighth House of joint resources, on the 20th, a conservative financial mindset is the wise choice. Confirm that bills are paid and hold off on important financial decisions until later in September. On the upside, this is a great time to organize important papers, review budgets, and research investments.

Rewarding Days
1, 5, 8, 9, 14, 17, 21, 22, 23

Challenging Days
4, 6, 12, 13, 18, 19, 20, 25, 27

Aquarius/September

Planetary Hotspots
Just when you thought the stresses and strains related to this summer's events were over, they return when the September 23 Full Moon in Aries activates Saturn in Libra and Pluto in Capricorn, your solar Ninth and Twelfth Houses. Use that week to resolve any lingering issues and to put the events behind you, while still observing the cautions of the past few months. Also strongly consider returning to school to complete a degree or to get the training you need for career advancement.

Planetary Lightspots
You're in high focus this month, thanks to Venus, which enters Scorpio, your solar Tenth House, on the 8th. Your popularity soars, and you can use this beneficial energy to gain supporters and promote your skills and talents. Midmonth could bring news related to your rising status. Cultivate decision-makers.

Relationships
Take time this month to touch base with people you haven't seen recently, as well as those at a distance. One of them could link you with an opportunity, but you'll need to get all the facts before moving ahead. Ultimately, you may decide that the best choice is to invest your energies elsewhere. The same is true if a reunion puts you in touch with someone from the past.

Money and Success
Between Mercury resuming direct motion in Virgo, your solar Eighth House, on the 12th and the New Moon in the same sign on the 8th, you could net a small windfall. It could come in the form of a raise or bonus for you, your mate, or both of you, or as an unexpected inheritance or lucky win. Be conservative, though, as there could be extra expenses this month.

Rewarding Days
2, 5, 6, 10, 14, 19, 20, 24, 28

Challenging Days
1, 3, 4, 9, 15, 16, 18, 23, 25, 30

Aquarius/October

Planetary Hotspots
Life is relatively calm this month, especially in comparison to the recent challenges. The exception is Venus, which turns retrograde in Scorpio, your solar Tenth House, on the 8th. This influence can put career plans on hold and make it tough for you to connect with the right people. Try anyway. Although you won't see immediate results, positive developments show up in November or December.

Planetary Lightspots
You need a break. Take it around the time of the October 7 New Moon in Libra, your solar Ninth House. Aim for a vacation trip, but if that's not feasible, give yourself the gift of a long weekend at a nearby destination. The change of scenery will refresh you and renew your spirit.

Relationships
Communication flows under the October 22 Full Moon in Aries, your solar Third House. You'll find it easy to connect with people then, and many will come to you for information. This planetary influence also encourages contact with relatives and neighbors, and you could be the ideal person to organize a project to improve or benefit your community.

Money and Success
Although retrograde Venus will put some career-related plans and projects on hold, others will take off, thanks to Mars in Scorpio through the 27th. Better yet, you'll be the driving force who encourages and motivates others to put forth their best effort. But be sure to ask for and listen to feedback. It will be easy to become so focused and impatient that you miss important details. Submit your résumé, but be patient. Success is more likely at year's end.

Rewarding Days
7, 8, 11, 15, 17, 25, 26, 29, 31

Challenging Days
2, 3, 4, 6, 13, 20, 22, 23, 28

Aquarius/November

Planetary Hotspots
November perks along with only minor challenges associated with your career and friends. Although it will be difficult to avoid a clash at work midmonth, you can rise above it by keeping your goals in sight. The same applies if a friend disappoints you. Try not to get upset, and instead view these events as excellent lessons in human nature that will serve you well in the future.

Planetary Lightspots
You're in tune with the domestic scene as the November 21 Full Moon in Taurus brightens your solar Fourth House. That's reason enough to plan a relaxing Thanksgiving with close friends and family, or to host an open house that weekend.

Relationships
Relationships get back on track after Venus turns direct on the 18th, and you can easily smooth over any recent difficulties with relatives. This month's best relationship influence, however, is Mars in Sagittarius, your solar Eleventh House. That's great for holiday socializing, the toughest part of which may be seeing everyone in your ever-widening circle. Accept every invitation that comes your way and see each as an opportunity to connect with people who can advance your career and personal aims. You never know who might be your lucky charm.

Money and Success
Your career life is on the upswing, thanks to the November 6 New Moon in Scorpio, your solar Tenth House. Work it to advantage and you could earn a nice raise or promotion. If you're job-hunting, the lunar influence could trigger a terrific and lucrative offer. Before you accept, however, be sure promises are realistic.

Rewarding Days
1, 4, 7, 8, 12, 13, 22, 26

Challenging Days
3, 9, 17, 18, 23, 24, 27, 30

Aquarius/December

Planetary Hotspots
December brings another retrograde Mercury cycle, beginning on the 10th in Capricorn, your solar Twelfth House, when you should take extra precautions to prevent a cold or flu. The emphasis shifts as Mercury retreats into Sagittarius, your solar Eleventh House of friendship, on the 18th. Then you'll need to confirm dates, places, and times so you don't miss out on a social event. A friend could try to mislead you this month, hoping to gain your sympathy and possibly your money. Don't hesitate to say no.

Planetary Lightspots
Although socializing will be a main focus this month, you'll also cherish time alone once Mars enters Capricorn, your solar Twelfth House, on the 7th. Take it without apology, and accept only those invitations that appeal to you. This is your time of the year to look within, to re-center before the Sun enters your sign in January.

Relationships
December features fun, friendship, romance, and socializing as the New Moon on the 5th and the Full Moon (lunar eclipse) on the 21st accent your solar Fifth/Eleventh House axis. You'll also feel especially strong ties to your children, who will want more of your time and attention now. The lunar energy will trigger a whirlwind romance for some, but if a relationship isn't working out as you hoped it would, now might be the time to move on.

Money and Success
Your career star continues to rise, thanks to Venus in Scorpio, your solar Tenth House, all month. Listen closely and get all the facts you can in early December when the grapevine will be active. What you hear may amount to nothing, or be just the opportunity you're looking for.

Rewarding Days
1, 3, 4, 5, 8, 10, 19, 23, 29

Challenging Days
7, 9, 14, 15, 17, 20, 21, 27, 28

Aquarius Action Table

These dates reflect the best—but not the only—times for success and ease in these activities, according to your Sun sign.

	JAN	FEB	MAR	APR	MAY	JUN	JUL	AUG	SEP	OCT	NOV	DEC
Move	23, 24	19, 20	16-31	3-16	12-19			2, 3				
Start a class		17	16-31									
Join a club								17, 18		11, 12	9-30	1-6
Ask for a raise		15, 16, 24, 25	4, 5									13
Look for work		24							30	1, 2		
Get pro advice			25, 26			15, 16	12, 13	9, 10			26, 27	
Get a loan		1						1-5				
See a doctor		9, 24	9						16, 17	1		
Start a diet								7, 8	3, 4, 30	1	28, 29	
End relationship	29, 30								5, 6	2-4	26, 27	
Buy clothes	25, 26		21, 22		14, 15	11-19			2		23	
Get a makeover	21-31	1-13	11									
New romance						12			2, 28		23	
Vacation								7-31		4-19		

Pisces

The Fish
February 19 to March 20

♓

Element: Water
Quality: Mutable
Polarity: Yin/feminine
Planetary Ruler: Neptune
Meditation: I successfully navigate my emotions
Gemstone: Aquamarine
Power Stones: Amethyst, bloodstone, tourmaline
Key Phrase: I believe
Glyph: Two fish, swimming in opposite directions

Anatomy: Feet, lymphatic system
Color: Sea green, violet
Animal: Fish, sea mammals
Myths/Legends: Aphrodite, Buddha, Jesus of Nazareth
House: Twelfth
Opposite Sign: Virgo
Flower: Water lily
Key Word: Transcendence

Your Strengths and Challenges

You can adapt to almost any situation or environment, whether to respond to the thoughts of others or to simply "disappear" when you don't want to attract attention. Intuition, or possibly psychic ability, helps you sense the prevailing mood and how people feel. But this ability can make you susceptible to environmental conditions, both positive and negative, so be aware and train yourself to consciously tune out potentially upsetting vibrations.

With Neptune as your ruling planet, you're sensitive and compassionate, and your heart goes out to those in need. These admirable qualities can at times encourage you to sacrifice your needs to help others who touch your sympathetic nature. That's not always the best choice, however. Sometimes the wisest decision is the toughest one: let others fight their own battles and learn the hard way.

Pisces is one of the most creative signs, and you probably use this talent every day even if you're not aware of it. Art, dancing, crafts, and writing are only a few possibilities, and your niche might be ideas and unique solutions.

Although you usually appear serene, your moods fluctuate and you have a tendency to brood despite your strong faith and spirituality. Try meditation or music, both of which can lift your spirits and help you re-center. Time alone to fill your mind with positive thoughts can also be beneficial.

Your Relationships

A select group of friends is one of the foundations of your life. Few in number, some are soul mates and lifelong friends, while others enter your life so you can learn from them. Quiet evenings with these people are your choice over meet-and-greet social events. You also have a wide circle of acquaintances, most of whom are in some way tied to your career life.

You're a romantic who remembers every special event. You also dislike being alone and need the peace of mind and happiness that comes with commitment. When dating, however, you can hang on long after you should part ways. You know this in your heart, but the security of a relationship often wins. Your ideal mate could be a Virgo, your opposite sign, and romance with one of the other two earth signs, Capricorn and Taurus, can also balance your sensitivity. You could make a match with one of the water signs—Cancer,

Scorpio, and Pisces—with whom you have much in common. But life with a Gemini or Sagittarius could be too unpredictable for you.

You're much closer to your siblings than you are to your family as whole. Siblings give you a sense of security, and time with them is relaxing and as comfortable as your favorite pair of jeans. You could have a similar relationship with an aunt, uncle, or cousin. Many moves during childhood or an ever-changing family life is the motivation behind your desire for close ties with your own children, as well as the urge to protect them from life's disappointing moments. Although that's natural for any parent, try to remember that children learn from these events, which will make them stronger, more independent adults.

Your Career and Money

Your best career choice is one that offers ongoing opportunities to expand your knowledge and that doesn't keep you behind a desk. It can be a challenge to find your career niche, mostly because you're interested in many fields. The secret is to combine those interests into one career rather than try this one and that one in hopes that you'll strike gold. In your daily work you need a job environment that's upscale, with high-quality tools and technology (and furnishings if you're in an office). Praise and recognition spur you on, and you need leadership opportunities. Being a big fish in a small pond could be your ideal situation. Your partner may be your best asset in money matters, which can be a challenge for you. Earning power is high, but so is spending; using credit wisely can be even more of a challenge. Listen to your mate and work together to develop a precise budget that includes savings, long-term investments, and spending guidelines. Save for big-ticket items and pay cash whenever possible.

Your Lighter Side

You have a unique ability to adapt to your environment in almost any situation, which gives you the edge when interacting with a wide variety of people in your business and personal lives. With this talent comes a reputation for charming people skills, and a knack for putting people at ease.

Affirmation for the Year

I'm ready for new adventures!

The Year Ahead for Pisces

This is a lucky year for you, thanks to Jupiter in Pisces—an event that occurs once every twelve years, bringing with it good fortune and optimistic enthusiasm. With the luck factor on your side, life will go your way more often than not. But Jupiter is also the planet of expansion, so it can represent too much of a good thing. Stick to a healthy diet and regular exercise to avoid adding extra pounds. You'll also want to keep events and opportunities in perspective because Jupiter can promise far more than it delivers and sometimes turn a minor mess into a much bigger one.

Jupiter transiting your sign is also the beginning of a new cycle of personal growth. All that you learn about yourself this year will serve you well in the next eleven. This is about finding a new personal niche and then defining how best to use that in the outer world to achieve career rewards and elevate your status. Snap up your chance to shine!

Jupiter briefly visits Aries, your solar Second House, June 5–September 7. Cross your fingers and think abundance because Jupiter here can increase your bank balance by triggering a raise, bonus, or windfall. But it will be tempting to be a big spender. Curb the urge to splurge and instead save your gains. The extra cash may come in handy for unexpected expenses. This is your opportunity to learn all you can about financial planning and money management. Then put all those principles to work for you.

Saturn begins and ends the year in Libra, returning to Virgo, your solar Seventh House, for its final transit of that sign April 7–July 20. You've undoubtedly experienced the many facets of relationships since Saturn entered Virgo in 2007. Some ties have strengthened, while others have left you wondering why you were ever close. You've probably also purposely distanced yourself from some people who no longer fit your lifestyle, interests, or activities. This final Saturn in Virgo transit will bring more of the same plus an opportunity to learn from a friend, mentor, or partner.

You may also decide that a close relationship has run its course, having done all you could in the past few years to explore available options. Only you can know the answer to this question as well as the relationship's future. In any case, this final Virgo transit will be a learning experience in human relations that will help you better define what you need, want, and expect in a partnership.

Joint resources are the focus of Saturn in Libra, your solar Eighth House. This transit emphasizes your partner's income, insurance, long-term investments, inheritance, and lenders. At least some of these areas of life will require attention during the next few years.

One of the best uses of Saturn in the Eighth House is to pay off any accumulated debt and to refrain from incurring more. If debt is an issue, set plans in motion to reduce credit card and other consumer debt to zero by 2012 (or sooner), when Saturn will move on to Scorpio. Pay cash, create savings, and develop a new habit of living beneath—not beyond—your means. Saturn will reward you with financial security and a higher credit score.

This is even more important because total family income could decrease at some point. The same is true of interest earned and return on investments. Take a close look at your retirement account and consider whether you should switch to a more conservative portfolio after researching options or consulting a financial professional.

An inheritance is possible as Saturn transits the Eighth House, possibly from a friend. But be cautious if a relative or friend asks for a loan or seeks your help as a cosigner. The odds are you won't be repaid and you could end up with full responsibility for a debt. This could occur in January or August when Saturn will clash with Pluto in Capricorn, your solar Eleventh House. You also may have to deal with insurance-related matters during this Saturn transit. Take time now to get updated comparison quotes, and be sure property is fully covered. Do the same with health insurance.

Like Jupiter and Saturn, Uranus spends time in two signs as it begins its transition from Pisces to Aries.

Personal change continues to be at the forefront with Uranus in your sign much of the year. If you haven't yet completed that process, now is the time. Or, you might be ready for round two in order to better reflect your growth as an individual since Uranus entered Pisces in 2003. Outwardly, this might take the form of a healthier diet and regular exercise so you're at your very best every day.

Uranus is equally active on a subconscious level, urging you to explore what motivates you as well as what holds you back. This developmental process complements the outer one by helping you to gain a new can-do attitude of self-confidence. Enhance your unique individuality as you take yet another step toward the future in your journey to become all that you can be.

Uranus will be in Aries, your solar Second House, May 27–August 12. This transit, like Saturn in Libra, emphasizes money matters, specifically your income, spending habits, credit, and debt. Take a close look at finances from a personal perspective and let Uranus motivate you to make necessary changes. What you do personally—decisions, actions, and follow-through—will affect the outcome of the Eighth House Saturn transit. So use your initiative to examine spending by closely tracking every expenditure for a month. You'll find the results both surprising and enlightening. Then put yourself on a budget, use credit sparingly (if at all), deal with debt, and commit to adding a healthy percentage of income to savings.

The Second House also represents what you value, from security to possessions to your skills and talents. Take time to examine this side of you and your life because the answers will reinforce new financial attitudes and motivate you to stick with your self-designed program. And while you're at it, tackle household clutter for both practical and symbolic reasons. Everything you toss out helps to clear the energy and prepare you for a thriftier lifestyle. You could even net some extra cash by holding a yard sale or taking unwanted items to consignment shops.

Neptune, your ruling planet, continues its long transit through Aquarius, the sign it entered in 1995. During these years with Neptune in your solar Twelfth House you've probably had periods where you preferred time alone to socializing and when your sixth sense was particularly active. This will be true this year as well, along with the desire to donate your time and talents to a good cause.

This year Neptune can create the illusion that all is well with a close relationship or prompt you to avoid the obvious. But try not to let your imagination run away with you even if you would rather avoid reality, and try to remember that Neptune is also the planet of faith and spirituality. Have faith in yourself and your own identity, strengths, and talents.

Pluto in Capricorn, your solar Eleventh House, will have a powerful influence on friendships during its long transit of this sign, which lasts until 2024. Friends will change your life and you theirs, and you'll attract new people into your life who are influential in the world at large. Many will appear to help you achieve a specific goal or fulfill a lifelong wish. But there will be difficulties with some friends, and others may try to use you for their own ends. Be particularly cautious of anyone who even subtly tries to manipulate you.

You could rise to a powerful leadership position in a club or organization during these years. Here, again, some caution is necessary. Keep things in perspective rather than letting Pluto's reputation for obsession turn what can be a positive experience into an obsessive one. On another level, you'll need to remember to lead rather than dictate when involved in any group activity, whether in your personal or professional life. Get a head start by polishing and practicing your leadership skills beforehand.

What This Year's Eclipses Mean for You

There are two solar and two lunar eclipses in 2010, three of which are in your solar Fifth or Eleventh Houses, highlighting romance, friendship, creativity, children, and socializing. The fourth is in your solar Fourth House of home and family. Each is in effect for approximately six months.

The January 15 solar eclipse in Capricorn, your solar Eleventh House, may be the best of the year because it's aligned with Venus, also in Capricorn, and Uranus in your sign. That's a terrific combination for networking, widening your circle of friends and acquaintances, and involvement in group activities. A friend could be your link to a wonderful opportunity or introduce you to someone who can open doors. This is also a great time to meet new people through a professional or charitable organization.

The June 26 lunar eclipse in Capricorn isn't nearly as positive. It's aligned with Pluto in Capricorn and Mercury in Cancer (Capricorn's opposite sign), as well as Jupiter and Uranus in Aries. What was upbeat in January regarding friends and groups could become tension-filled, complete with power plays. Step away if your values clash with those of a friend or group, and don't feel obligated to contribute money, no matter how good the cause. You also should steer clear of any fundraising activities. But this eclipse could also trigger a terrific opportunity; just be sure to ask questions and base your decision on facts, not promises.

July 11 brings a solar eclipse in Cancer, your solar Fifth House. This is a wonderful influence for parents and children, and for singles looking for love. Its alignment with Mars in Virgo could trigger a minor windfall, but this influence is even better for sports and other outdoor activities. Take tennis or golf lessons, join a gym or sports league, or get involved in your children's activities as a coach. Or nurture and express your creativity by learning a new hobby or returning to one from the past. And be sure to plan plenty of time for socializing with friends.

The year concludes with the December 21 lunar eclipse in Gemini, your solar Fourth House of home and family. The domestic scene will capture your interest and you'll be motivated to begin (or finish) home decorating and improvement projects. But this eclipse also aligns with Mercury in Sagittarius, your solar Tenth House of career, so it will be a balancing act to devote necessary time to family and your worldly responsibilities. Resist the urge if a home-based business has sudden appeal, especially if it would require a sizeable financial commitment. Keep your day job and develop a sideline during the evenings and weekends.

Saturn

If you were born between March 17 and 20, Saturn will contact your Sun from Virgo, your solar Seventh House, between April 7 and July 20. A close relationship will reach a turning point during this time. Whether it ends or continues is a choice for you and the other person, but be sure you've explored every possible avenue before cutting ties. That will help satisfy the learning experience that Saturn represents. Otherwise, you can expect to repeat the lesson in seven or fourteen years. Saturn is also the ultimate karmic planet, so let your actions reflect this planet's higher vibration.

Your energy level may be low during this transit. Plan ahead to get plenty of sleep and do your best to set aside time to relax every day. Let family members do their share rather than pushing yourself to the max.

A friend may disappoint you or you might see another, less attractive side in someone you've been close to for many years. It's also possible someone from the past could reappear in your life. Remember the negatives as well as the positives before you renew the relationship. If, however, there are people and issues from the past you need to resolve, now is the time to put them to rest. Then you'll be free to move onward and upward to reap the Saturn rewards that will come in about five years.

If you were born between February 19 and March 6, Saturn will contact your Sun from Libra, your solar Eighth House of joint resources. But only **if your birthday is between February 19 and 23** will you experience the full effect of this Saturn transit between January 1 and April 6, and again from July 20 through early August. **If you were born after March 23**, Saturn will contact your Sun for only a week or two between mid-August and year's end.

Finances will require more than a little of your attention during this transit, when you could unfortunately see family income decrease. Whether or not that occurs, it's time to get serious about getting your finances in order, including debt reduction, building up savings and retirement funds, and generally adopting a thrifty attitude and budget. As in all things Saturn, this too is a learning experience.

You also want to be sure your property and possessions are fully insured so that there are no difficulties should a claim be necessary. It's possible there could be changes or higher premiums for health insurance, so be sure to read all the fine print before selecting or revising coverage.

During this time you could hear news about an inheritance, or an old financial matter could be settled. If you're owed money, you may be able to collect it now. Conversely, someone could ask you to settle an old debt. Remember, Saturn smiles when rules and responsibilities are followed and met.

Uranus

If you were born between March 14 and 20, Uranus in Pisces will merge its energy with your Sun between January 1 and May 26, and between August 13 and December 31. This can be one of the most exciting years of your life because you're open to personal change, which is Uranus's specialty. Look to the future, create a new direction for yourself, and get started. You might want to hold off until later in the year, however, to make any major lifestyle change. Action taken under a Uranus transit is rarely reversible, so first be sure you're going where you want to go.

Individuality and freedom are keynotes of this transit, which encourages you to set a new norm for yourself. That can take many forms, and ultimately you're the only one who can redefine yourself and your place in the world.

If Saturn in Virgo will also contact your Sun, significant changes in a close relationship or partnership may be on the horizon this spring or summer. But it's not so much about the relationship as it is about you. The changes you're undergoing will simply impact the way in which you view the other person and your commitment. If your partner is open to the new you, you may be able to redesign your relationship to suit your changing needs. If not, you have a decision to make; but don't act in haste no matter how you feel.

If you were born February 19 or 20, Uranus in Aries will contact your Sun from your solar Eighth House of joint resources. You, among all born under your sign, should overhaul your finances in some way even if you feel everything is on track. Analyze spending, saving, and debt, and check your (and the family's) credit reports for errors.

Your income may shift up or down, and a windfall is nearly as likely as unexpected expenses. If you do have a gain, save before you spend. An inheritance or legal or insurance settlement could be more or less than expected. Above all, this is not the time to put funds at risk, despite what someone may advise you. Commit to a conservative path.

Uranus, planet of the unexpected, could lead you to a lucky find on the street, in a thrift shop, at a yard sale, or in your basement. But you also could lose a valuable item, so always protect possessions when you're out and about. Do the same with sensitive financial information, and be cautious when banking or using a credit card online.

Neptune

If you were born between March 14 and 20, Neptune will contact your Sun from Aquarius, your solar Twelfth House. Your sixth sense will be more than usually active this year, and hunches can be completely on target. You'll want to back them up with facts, however, because Neptune is known for confusion and illusion. It's also an inspirational planet, so look within for the motivation you need to complete any task. Have faith in yourself and your abilities.

You'll be drawn to those in need and may get involved in a charitable organization. It's also possible an ailing friend or relative will need your help at some point during the year. Do what you can, but don't put your life on hold; share the load with others.

Dreams will be more vivid this year, and some could be prophetic. Put pencil and paper next to your bed to record your impressions and look for emerging themes. They can help you understand the true source of your worries as well as what you truly wish and hope for.

Pluto
If you were born between February 21 and 24, Pluto will contact your Sun from Capricorn, your solar Eleventh House. This is your sector of groups and friendship, both of which will be prominent in your life this year. Step up and take the lead in any teamwork setting, motivating others to do and be their best and guiding them to consensus. But keep yourself in check because it will be tough at times to let the discussion and process unfold as it will and not necessarily according to your plans or schedule.

Tap into Pluto's willpower and make it your own. Transforming almost any part of yourself will be much easier now because you'll have the determination of this planet on your side. But also know when to quit because it will be easy to push yourself (and others) too hard to achieve a goal.

A friend could transform your life this year, or you could do the same for someone else. And anyone new who enters your sphere has the potential to help you succeed through networking. Get to know those who can advance your status, and also review, revise, or establish solid career and financial goals.

Jupiter-Saturn-Uranus-Pluto
If you were born between February 19 and 22, events will occur unexpectedly and unfold rapidly when this four-planet alignment contacts your Sun this summer. Although it's tough to forecast anything when Uranus is involved, the lineup encompasses your solar Second House (Jupiter and Uranus in Aries), Eighth House (Saturn in Libra), and Eleventh House (Pluto in Capricorn).

The financial theme is prominent with planets in both the Second and Eighth Houses, so you'll need to be cautious with money matters. Saturn restricts, Jupiter expands, and Uranus adds the element of surprise. That could just as easily trigger a nice windfall as a major expense, or possibly both. There may be difficulties concerning other people's money (debt, inheritance, insurance, legal affairs), or you could cash in on a group lottery purchase with friends, represented by Pluto in Capricorn. This alignment could also bring a change in income, either up or down, and you should think carefully before getting involved in anyone's (or a group's) financial affairs, or forming a partnership that requires an investment from you. Other difficulties could arise in a group or organization that require you to make tough choices.

Pisces/January

Planetary Hotspots
Mars is retrograde all month (and through mid-March) in Leo, your solar Sixth House. That's enough to make delays and frustration an all-too-often occurrence at work. The energy will peak the week of the January 30 Full Moon in Leo, so be prepared to keep cool. Tense talks are best handled in private. Even so, say and write only what you're willing to share with others.

Planetary Lightspots
January 17 is a date to remember. That's when bountiful Jupiter enters your sign, bringing with it the luck potential this planet is noted for. You can wait for good fortune to come to you, but a better choice is to make your own. Seek opportunities and follow through on those that pop up, no matter how improbable they may seem.

Relationships
The January 15 New Moon (solar eclipse) in Aquarius spotlights your solar Eleventh House of friendship—a real plus for your social life. It's also an excellent influence for networking, and someone you meet midmonth could be a lucky connection. Consider joining a club or organization to widen your circle, or organize a special interest group with coworkers to master a new skill that ultimately could benefit your career.

Money and Success
You could feel a financial pinch as Saturn in Libra, your solar Eighth House of joint resources, clashes with Pluto in Capricorn. A friend may ask for a loan or another financial favor. Think carefully. Chances are, you won't see your hard-earned money again. Also take time to seriously review family finances and related goals. You have the determination and willpower now to live within a strict budget, if necessary, and to take action to become debt-free.

Rewarding Days
1, 5, 14, 17, 18, 19, 24, 28

Challenging Days
4, 6, 13, 16, 20, 23, 27, 31

Pisces/February

Planetary Hotspots
This month's hotspot can also be a lightspot, depending upon your choices. The February 3 New Moon in Aquarius encourages you to take a close look at your lifestyle. Healthy diet? Rest? Relaxation? Exercise? Habits? Be honest with yourself and make needed changes. All these things can boost your immune system and help prevent a cold or flu, which you're more susceptible to now.

Planetary Lightspots
Share your wishes with the universe after Venus enters your sign on the 11th. You can attract exactly what you want if you truly believe you're deserving. Results could come as soon as the days around the 16th, when Venus joins Jupiter in your sign, creating a window of maximum luck potential. Go for it!

Relationships
The February 28 Full Moon in Virgo, your solar Seventh House, spotlights close relationships—friends, partner, family. One or more of these relationships will reach a turning point, upward or downward, closeness or distance. A serious dating relationship could move to the next level or come to an end. Each contact, however, will teach you a little more about human nature and enhance your people skills.

Money and Success
Work-related frustrations continue with Mars still retrograde in Leo, your solar Sixth House. A midmonth disagreement could push your limits, but try to stay calm and keep your goals and income potential in sight. Do that and you could be rewarded in late March.

Rewarding Days
1, 3, 10, 11, 14, 15, 16, 24, 25

Challenging Days
2, 4, 8, 9, 12, 17, 19, 21, 23

Pisces/March

Planetary Hotspots
Although expenses will rise under difficult planetary alignments this month, favorable contacts will help bridge the gap if you're careful and tighten your budget. Deal with challenges as they arise rather than avoiding reality, and take the time to check credit reports. Curb your generosity when socializing and let friends pay their own way.

Planetary Lightspots
March brings the start of your new solar year with the New Moon in your sign on the 15th. Even better, it's aligned with Mercury and Uranus, also in Pisces, giving you fresh insights into yourself and your talents. Use that information to set yourself on a solid path of achievement for the next twelve months. You have a lot going for you. Make the most of it!

Relationships
With three planets (Sun, Venus, and Mercury) in Pisces part of the month, you'll have an extra level of magnetic charm. A chance encounter could connect you with a new friend, love interest, or networking contact. But later in March there could be some tense moments with a pal, possibly involving money. Stick to your values and preserve your resources.

Money and Success
Mars in Leo, your solar Sixth House, finally resumes direct motion on the 10th. Work projects and plans put on hold will gradually regain momentum and you'll soon be back up to speed. Solo work will be more productive than teamwork, which will try your patience this month. Nevertheless, join in, because cheerful participation will reinforce your position.

Rewarding Days
4, 5, 9, 10, 13, 14, 15, 19, 27, 28

Challenging Days
2, 8, 11, 16, 18, 20, 23, 25, 29

 # Pisces/April

Planetary Hotspots
Saturn returns to Virgo, your solar Seventh House, on the 7th and aligns with Uranus in Pisces on the 26th. This signals stress and strain in a relationship when the desire for change clashes with the status quo. It may be enough to push things to the breaking point. Your other option is compromise. Try to appreciate the other person's viewpoint and learn from it—whether you ultimately maintain or dissolve the tie.

Planetary Lightspots
Although it will be the 25th before Venus enters Gemini, your solar Fourth House, when that happens you'll have renewed enthusiasm for the domestic scene. Start thinking about new décor. With the Full Moon in Scorpio, your solar Ninth House, on the 28th, you can get the knowledge you need for do-it-yourself projects and beauty on a budget.

Relationships
Relationships are prone to mix-ups and misunderstandings from the 17th on, when Mercury turns retrograde in Taurus, your solar Third House. Choose your words with care, and confirm dates and destinations. You should take your vehicle in for routine maintenance well before Mercury's retrograde period begins, and be sure to check the battery and spare tire.

Money and Success
Finances will be in positive territory, thanks to the April 14 New Moon in Aries, your solar Second House. Keep spending in check, though, by shopping sales after Mercury turns retrograde, when you could get some terrific deals. Use your sixth sense to look for mismarked items—wrong size, wrong price—and check all racks, not just those with your size.

Rewarding Days
1, 6, 7, 10, 11, 16, 20, 27, 29

Challenging Days
2, 4, 5, 12, 15, 19, 22, 23, 25, 28

Pisces/May

Planetary Hotspots
Uranus enters Aries, your solar Second House, on the 27th for a three-month visit. Although you're unlikely to experience the financial ups and downs of this influence in May, Uranus will be more active in June, July, and August. Save money now. You may need it to cover increased expenses later this summer. Your best bet is to shop only for necessities; this way, you can minimize impulse buys, which will tempt you now.

Planetary Lightspots
Take advantage of warmer temperatures and Venus in Gemini, your solar Fourth House, to get your yard in shape or to plant a patio flower or vegetable garden. Then move inside and tackle repairs and touch-ups. Show off the results by hosting a get-together for friends or family later in May.

Relationships
Make the effort to clear up any recent misunderstandings after Mercury resumes direct motion in Taurus, your solar Third House, on the 11th. People will appreciate the gesture, and you can get a friendship or another close relationship back on solid ground. However, tension surrounds one friendship (or a dating relationship), near the 23rd, when it will be tough to find a compromise. Try anyway and keep an open mind.

Money and Success
Use your attention-getting power to impress decision-makers when the May 27 Full Moon in Sagittarius, your solar Tenth House, has you in the career spotlight. But be careful not to promise more than you can deliver. Otherwise, the very gains you're hoping and working for might evaporate. Balance enthusiasm with reality by keeping your big-picture goal in sight.

Rewarding Days
1, 3, 7, 12, 17, 21, 25, 30, 31

Challenging Days
2, 4, 9, 16, 18, 22, 23, 27, 29

 # Pisces/June

Planetary Hotspots
Tension rises on the home front, sparked by domestic and family issues. Disagreements over expenses and spending, children, relatives, and major purchases are possible, and it will be tough to balance everyone's needs and desires. You can make progress, however, by focusing on values and the practical realities. In-laws or another family member could aggravate the situation.

Planetary Lightspots
Look forward to the 14th, when Venus enters Leo, your solar Sixth House of daily work. You'll enjoy your job more than ever, along with added popularity that has your star shining. You'll also have luck on your side, and coworkers will pitch in when necessary. Be a team player and return the favor.

Relationships
Expect challenges with some friends and groups as the June 26 Full Moon (lunar eclipse) in Capricorn spotlights your solar Eleventh House. Difficult people will show their most difficult side. Stand your ground but skip the debate because it will be tough to change stubborn minds and fixed opinions. Ultimately, your best option may be to end the friendship or group association. Do as asked, however, if you see a potential clash developing in the workplace.

Money and Success
Cross your fingers for what could be a nice windfall, thanks to Jupiter's merger with Uranus in Pisces, your solar Second House. But this combination could just as easily trigger an unexpected expense. When Uranus is involved, almost anything is possible. Save if you gain, because this planetary energy will be active through August.

Rewarding Days
4, 8, 9, 13, 14, 18, 22, 27, 30

Challenging Days
3, 5, 6, 10, 12, 19, 24, 25, 26

Pisces/July

Planetary Hotspots
Finances could take a downturn as Saturn in Libra aligns with Uranus in Aries, your money sectors. (Saturn enters Libra on the 21st.) A drop in income or benefits is possible, along with credit issues. Be proactive with lenders rather than to let things slide, and try not to incur debt. Safeguard valuables, protect sensitive financial information, and check your credit reports for errors. There's also a chance you could collect a long overdue debt.

Planetary Lightspots
The July 11 New Moon (solar eclipse) in Cancer, your solar Fifth House, energizes your leisure-time sector, highlighting romance, creativity, socializing, and children. Fill your calendar with events and do it all! Plan an inexpensive family outing the second weekend of the month, or host a potluck at your place. The timing is also excellent for singles looking for someone new.

Relationships
Relationships get a much-needed positive boost after Venus enters Virgo, your solar Seventh House, on the 10th, followed by Mercury on the 27th. But the best news for this sector is Saturn's completion of its multi-year Virgo transit on the 20th. You'll be in sync with friends midmonth and also the end of July.

Money and Success
July 25 brings a Full Moon (lunar eclipse) in Aquarius, with the Sun in Leo, your solar Sixth House of daily work. This planetary energy has potential for a step up or a new position, but the competition will be stiff and salary may be less than expected. Weigh the pros and cons and give careful thought to the long-term as well as immediate gains, including the possibility for advancement. Be realistic.

Rewarding Days
1, 2, 5, 6, 7, 11, 14, 18, 24, 29

Challenging Days
3, 9, 10, 12, 16, 19, 22, 23, 30

Pisces/August

Planetary Hotspots
Money matters are front and center this month as the Jupiter-Saturn-Uranus-Pluto lineup reaches its peak. A change in income and increased expenses can accompany this planetary influence, and there could be financial difficulties involving a friend or group you're associated with. Make no loans, and don't cosign or let anyone use your credit. Be sure property and possessions are fully insured, and continue to minimize spending and use of credit.

Planetary Lightspots
The August 24 Full Moon shines brightly in your sign, adding a dash of pizzazz to your easygoing charm. That can be an especially strong advantage in connecting with the right people, who can help you network your way to success. Someone at a distance could be the most helpful of all.

Relationships
Relationships can be bumpy after Mercury turns retrograde in Virgo, your solar Seventh House, on the 20th. Clarify your thoughts, ideas, and feelings rather than letting others fill in the blanks for you. At the same time, though, listen to other viewpoints and blend them with yours. If possible, hold off until mid-October if you need to consult a professional for advice.

Money and Success
You'll be scrambling to keep up with everything at work this month. Stay organized and take things one at a time. The August 9 New Moon in Leo, your solar Sixth House, also favors job hunting if you're in the market for a new position. Send out résumés the following week and you could hear encouraging news by the end of September, if not before.

Rewarding Days
2, 3, 5, 7, 11, 14, 16, 21, 26, 29, 30

Challenging Days
4, 6, 10, 12, 13, 18, 19, 20, 27, 28

Pisces/September

Planetary Hotspots
Money matters are a hotspot again as the September 23 Full Moon in Aries activates your solar Second House, as well as Saturn in Libra and Pluto in Capricorn. Your partner's income could be affected, and you may have to deal with an insurance or legal matter. It's also wise to hesitate before trusting someone close to you with financial information. Safeguard possessions, which are easily lost now, and always catch a ride if you're out socializing.

Planetary Lightspots
Jupiter returns to your sign on the 8th, for the rest of the year. When it joins forces with Uranus on the 18th, that week could bring an exciting opportunity when you least expect it—just because you'll be in the right place at the right time, surrounded by good fortune.

Relationships
You'll be drawn to people and they to you when the September 8 New Moon in Virgo highlights your solar Seventh House of relationships. The timing is perfect for couples and commitment as well as upbeat get-togethers with your inner circle. And with Mercury resuming direct motion in Virgo on the 12th, you can easily resolve any misunderstandings that occurred in the previous three weeks. Be cautious, though, if you're in a teamwork situation or another group activity. Someone could try to undermine your position.

Money and Success
Venus will be in Scorpio, your solar Ninth House, from the 8th on, where it's joined by Mars on the 14th. Business travel is possible, and you can best use this time to take a short-term class or to get up to speed on a new technique or technology. Teamwork can benefit your career. Ask others to share their knowledge.

Rewarding Days
6, 8, 11, 12, 16, 17, 20, 22, 25

Challenging Days
1, 3, 4, 7, 9, 18, 23, 27, 30

Pisces/October

Planetary Hotspots
Answers are easy to come by this month with Mercury zipping through three signs—Virgo, Libra, and Scorpio. The question, however, is what and whom to believe. Be wary of anyone who pushes too hard, makes lofty promises, or tries to mask the facts with overly optimistic enthusiasm. Be skeptical and ask someone you trust for an objective opinion.

Planetary Lightspots
Mercury in Virgo, your solar Sixth House, connects with Jupiter and Uranus in Pisces the first two days of the month. Think luck and opportunity. Someone you know or someone new could be your lucky charm. Some singles will launch a sensational love-at-first-sight romance, and others will get a fabulous surprise from the one they love.

Relationships
Relationships in general will go into a distant mode when Venus turns retrograde on the 8th. People will pull back and be less supportive during the next six weeks. This is not the time for a wedding. Commitments made now will be short-lived, and a long-distance relationship is likely to dissolve. With Venus turning retrograde in Scorpio, your solar Ninth House, travel is prone to delays and lost luggage; postpone a trip if you can.

Money and Success
Money matters move into more positive territory as this month's New Moon (October 7) in Libra and Full Moon (October 22) in Aries highlight your financial sectors. Recent challenges can be overcome now, and finance-related talks and communication will benefit from Mercury in Libra October 3–19. Practical solutions are more easily found in early October.

Rewarding Days
1, 5, 8, 10, 18, 19, 24, 29

Challenging Days
2, 6, 12, 13, 14, 20, 21, 27, 28

Pisces/November

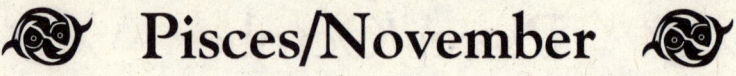

Planetary Hotspots
Life perks along with only minor bumps in your path. The possible exception is retrograde Venus, which slips back into Libra, your solar Eighth House, on the 7th before resuming forward motion on the 18th. Money may not flow quite as freely, and you should hold off on major purchases as well as shopping for clothing and gifts until then. But this is an excellent time to develop a holiday budget and to window-shop online.

Planetary Lightspots
Travel may be in the forecast under this month's New Moon in Scorpio, your solar Ninth House, on the 6th. Whether for business or pleasure, the getaway will refresh your soul and spirit. If time is short, aim for a long weekend near the November 21 Full Moon in Taurus, your sign of quick trips.

Relationships
Although it will be month's end before your holiday social life gets in gear, once Mercury enters Capricorn, your solar Eleventh House, invitations will begin to arrive. Make plans now to host a get-together at your place next month, when the December 21 New Moon in Gemini highlights your solar Fourth House of home and family.

Money and Success
Your career sector is in high focus with Mars in Sagittarius, your solar Tenth House, picking up the pace at work. The red planet is joined by the Sun and Mercury part of the month to give you a great opportunity to showcase your skills and talents. Watch for the chance to take the lead in a project or possibly to take on extra responsibilities. Either one could net you some extra income both this month and next.

Rewarding Days
4, 5, 6, 8, 14, 16, 19, 20, 26, 28

Challenging Days
1, 3, 7, 9, 15, 18, 23, 24, 27, 30

Pisces/December

Planetary Hotspots
Friendship is a strong theme this month with three planets in Capricorn, your solar Eleventh House. Easy alignments signal upbeat get-togethers in early December, but the reverse is true in mid-December and at month's end when personalities and egos will clash. Do yourself a favor: avoid the people you'd rather not see during that time frame, even if someone pressures you to join the crowd.

Planetary Lightspots
Home and family are the perfect match for the holidays, thanks to the December 21 Full Moon (lunar eclipse) in Gemini, your solar Fourth House. Invite a group to join you and yours, or travel to your hometown for a family get-together. Earlier in the month, consider hosting an event for friends and coworkers.

Relationships
Mercury in Capricorn slips into retrograde motion on the 10th, so be sure to confirm plans before you go. Take similar precautions on the job and double-check all work after Mercury retreats into Sagittarius, your solar Tenth House, on the 18th. Also choose your words with care, reread e-mail before you hit send, and clarify all work-related projects.

Money and Success
The month gets off to a beautiful career start under the December 5 New Moon in Sagittarius, your solar Tenth House. Set an ambitious agenda and show decision-makers you have what it takes. Rewards—or the promise of them in 2011—are possible. Finesse and persuasion will be required with coworkers and supervisors at month's end, however. Resist the urge to push others, which will result in a setback. Find a positive stress-reliever.

Rewarding Days
2, 3, 4, 5, 8, 10, 13, 17, 23, 29, 30

Challenging Days
7, 9, 14, 15, 20, 21, 22, 27, 28

Pisces Action Table

These dates reflect the best—but not the only—times for success and ease in these activities, according to your Sun sign.

	JAN	FEB	MAR	APR	MAY	JUN	JUL	AUG	SEP	OCT	NOV	DEC
Move	25, 26		21, 22			11–19						
Start a class	23			5–16	12, 13		7	2, 3				
Join a club	1–18			5, 6					16, 17			7, 8
Ask for a raise			16–18, 25, 26									
Look for work			25, 26			1–5, 15, 16	12, 13, 23–26	9, 10			26, 27	
Get pro advice		1, 2			21		14	2, 3, 11	16, 17			
Get a loan						19						
See a doctor		12	25, 26		5		12	13			24	
Start a diet									3, 4, 30, 31		26, 27	
End relationship		28	1							5, 6	28, 29	
Buy clothes		24, 25		20	17, 18	25–30	1–8		30	1		
Get a makeover		12–28	1–6	10					23			
New romance		24, 25				13, 26–30	1–11					
Vacation			4, 5							21–31	1–7	

Blank Horoscope

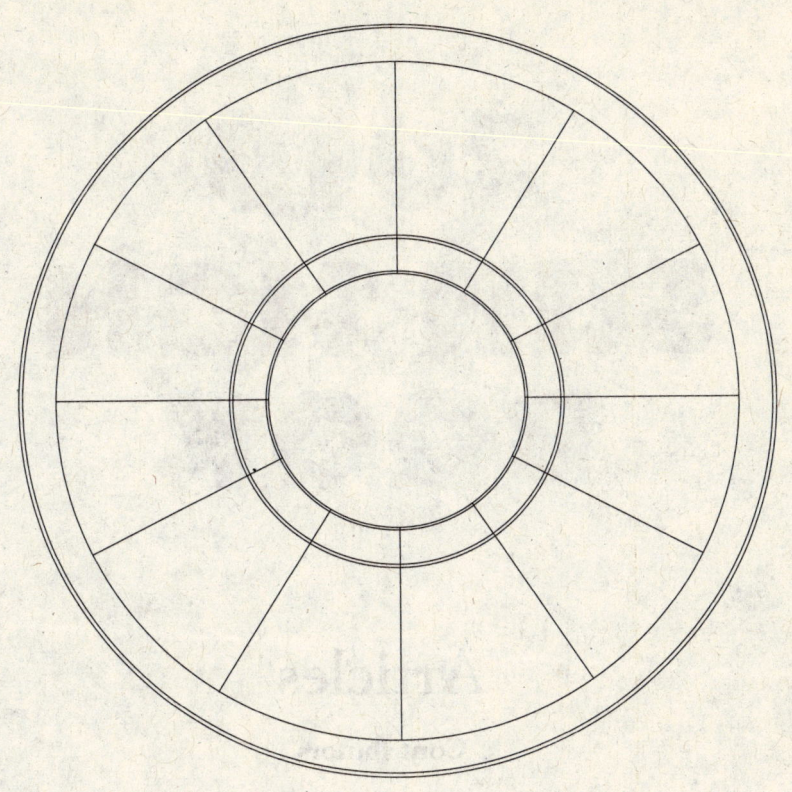

Articles

Contributors

Alice DeVille

Lesley Francis

Robin Ivy

Shakti Navran

Bruce Scofield

Kaye Shinker

The Mystery of Healing Gemstones

by Shakti Navran

Have you ever fallen in love at first sight with a piece of jewelry or a gemstone? Perhaps you couldn't get it out of your mind, or you dreamed about it. I believe many of us have had that experience, because jewelry and gemstones are so much more than just pretty accessories. There are different ways to approach jewelry. At the most basic level, you can simply allow yourself to be guided intuitively to the right piece for personal adornment. But if you would like to take advantage of the incredible balancing and healing properties of gemstones, you can apply the ancient science of astrology and its universal principles to select exactly the right stone.

This article provides an overview of the healing techniques I have studied and practiced over the past thirty years. If you are interested in knowing more, I discuss how astrology and the healing properties of gemstones can be combined to create powerful custom jewelry in my book *Jewelry & Gems for Self-Discovery* (Llewellyn, 2008).

Our Complex Body, Mind, and Soul System

Today, even conservative medical practitioners agree that the body and mind are intricately linked and inseparable. Spiritual practitioners also include the soul in the mix. Moreover, physicists tell us that everything in the universe consists of energy in various states—this is not a New Age idea at all, but a basic law of physics. So body, mind, and soul are different manifestations of the divine energy that created us. All run on energy, and that energy needs to be finely tuned for optimal performance.

In a well-tuned system, body, mind, and soul are aligned harmoniously and all is well. But much can go wrong in this complex system and, as a result, we may suffer from a great many physical, mental, and spiritual ailments. If one component of the system is out of sync, the other parts will be affected and the whole will be diminished.

Often, it is not helpful to treat only the external symptoms of illness, as doing so would be like turning off the check-engine light in your car without discovering why the light is on. The deeper cause of illness must be found. If someone is gaining weight even though nothing has changed about their nutritional and exercise regimen, the cause may be hormonal. And what is the cause of that? The endocrine system is impacted by physical, mental, and spiritual factors, all working together in unseen ways.

The knowledge that everything in the universe, including our body, mind, and spirit, consists of energy opens up many new healing modalities that work with our energy system. For example, EKGs, EEGs, and MRIs use energy for diagnosis. Alternative practitioners use new tools such as cold laser machines, magnets, and Rife computers to deliver various frequencies with the intention of restoring energy balance. Those technologies are based on knowledge that was known thousands of years ago in the ancient Indian and Chinese healing traditions. Homeopathy, acupuncture, and Ayurveda are proven ancient energy therapies.

The healing technique I would like to put in your hands is the exquisite beauty and energy of gemstones. The crystal structure of gemstones bears witness to the amazing creative and organizing forces of the universe, mirroring the much larger cosmic forces of the heavenly bodies. I've formed an intimate relationship with gemstones in the thirty years that I've worked as a jeweler. My life's work is to come to know through study and practice how to use the healing power of these mysterious gifts of the universe. I believe that by carefully choosing the gemstone that corresponds to imbalances and life challenges, it is possible to experience deeper joy, balance, and peace.

Jewelry and Gems for Healing and Self-discovery
Most of us come to a point at some time in our lives when we want to transcend the limitations and pain of the past. When this impulse arises out of a particularly difficult life challenge, it pushes us to seek understanding and a new direction. In astrological terms, this period may coincide with what is called a "hard transit"—when a transiting planet forms a hard aspect (conjunction, square, or opposition) to a planet in the natal chart—that causes frictional energy between the planets. A hard transit is a sign of a major change in progress and people often consult an astrologer at those times to gain insight into the causes and meanings of the upheaval they feel. The individual's birth chart allows us to take a step back and gain an overview of their internal design, qualities, difficulties, and capabilities.

The birth chart and its current transits provide guidance and insight into conflicting feelings and different aspects of the personality and, all of a sudden, it is possible to understand what is at the root of unsettled feelings. The birth chart also offers a fascinating view of the thematic challenges and trials that may recur throughout one's life. These are the areas where the most work is needed, and it is important to come to an understanding of the principles underlying thematic challenges.

For example, difficulty with authority figures is, in essence, a struggle to grow into one's own strength and power. Challenges with authority figures may appear in the birth chart as a Sun-Pluto opposition. The Sun represents ego identification—how you see and express yourself in the world—and Pluto forces the self-identity to grow through a process of deep transformation by questioning the principles that he intersects within the chart. This aspect will play out as friction with authority figures—the boss, parents, or other people who are viewed as powerful.

By understanding the underlying principle of the continuous struggle with controlling and powerful people, it is possible to transform habitual responses. When we accept our personal power and our responsibility to exercise it, we cease to be victims and conflicts with authority dissolve. This is an evolutionary process we all have to go through to some extent, not only rebellious teenagers.

Likewise, many of us have experienced the painful end of a relationship or marriage. It's a time of feeling very insecure, of not wanting to be alone, and of wondering if it will be possible to survive alone. The desire to return to the security and intimacy of a couple is strong because so much freedom has not been a part of life. At the same time, there is relief that the turmoil of the relationship is over, that freedom and positive changes are now an option.

Once these main conflict themes are identified, it is possible to choose gemstones with specific characteristics to help balance those conflicts. This is referred to as a stone's "signature," meaning the different planetary principles that compose the individual healing mode of a particular gemstone.

To help individuals navigate their life path I have created what I call "jewelry for the soul," highly personalized pieces that match the needs revealed in the birth chart with the qualities of specific gemstones. When creating jewelry for the soul, I consider the specific vibration and frequency of the gemstone, the individual's conscious empowerment through understanding the issues that are to be addressed, and esthetic qualities for personal enjoyment of a jewelry piece.

Universal Principles Are at Work Everywhere

Astrological principles manifest at all levels of creation, from the largest bodies in the cosmos to the tiniest particles of atoms—all are part of the same network of universal energy. Based on the hermetic principle, "As above, so below," those universal principles allow us to read the analogue relations among the planets and extrapolate those principles to things here on Earth, to events in our lives, as well as to our natural environment. The universal principles that govern the planets are also seen on the much smaller level of gemstones, crystals, and metals.

Let's take the example of hematite to illustrate. The chemical composition of hematite is Fe_2O_3, or ferric oxide. It is a silvery-gray metallic to black opaque stone that bleeds red into the water when cut and polished. We know that the color red and iron are associated with Mars, the red planet. Mars is the warrior and conqueror and represents our male, physical, and sexual energy; our life force, our aggression, impulses, will, discernment, and courage. So hematite will be helpful for all Mars-related matters. It can help strengthen the life force, balance aggression and sexual energy, and assist us with our will, courage, and discernment. If fatigue or listlessness or lack of direction is a problem, using a gemstone with Mars qualities can be supportive and help to realign personal energy flow.

Every internal or physical process you find yourself in is represented by some astrological planetary configuration. Understanding the planetary representation gives you the conscious choice of how you want that principle to play out in your life and how you can support the process you are in with gemstones.

Do Gemstones Radiate Healing Energy?

Thirty years ago, when I started inquiring into the possibility that gemstones do have some kind of healing effect and radiation, I participated in a seminar devoted to experiencing gemstones and crystals. What caught my attention in this seminar was a meditation with a rubellite (pink tourmaline). I had the most incredible, sweet, loving experience you can imagine. I felt pink, loving light pouring the most intense joy and ecstasy into every cell of my being. I was that unconditional love and sweetness. I was that compassionate heart. That was the first time I had what I would call a "cosmic experience."

I have always been a firm believer in valuing my own experience over book knowledge or what other people say. That experience got me hooked and started me on my journey of spiritual discovery, and it led me to many teachers, including shamans, Western scientists, East Indians, Sufis, and meditation teachers.

Some people I met started out as healers or spiritual teachers and then began to use crystals to enhance transmission of their own healing energy. Others used crystals for meditation or laid them in healing patterns on the body. What I learned from these teachers allowed me to deepen my knowledge of healing and my connection to my own inner being. And later, as I was passing on my own knowledge to my clients and students, I learned a great deal through their experiences. It became clear to me that using gemstones was intensifying my inner journey. My spiritual path is the center of my life, so I wanted to learn how to use gemstones to support my spiritual journey; to help me grow and heal on subtle levels.

The question remains for the curious: just how do the gemstones bring about healing? Science tells us that all matter is energy in one form or another. That energy is emitted in the form of waves that vibrate at a certain frequency. Because nature has created gemstones with a very specific and balanced crystal structure, they are perfect, harmonious compositions that radiate balanced energy into our system. We know this intuitively and, thus, we are drawn to them for their healing and balancing properties. Most people are able to sense this radiation of energy. They might say, "This is such a beautiful stone!" and be drawn to it because of the positive energy they feel.

Gemstones also radiate colored light frequencies into our own energy field. Their translucent quality is a bridge between the material and the spiritual realms. Because gemstones represent the highest form of radiant light frequencies in physical matter, they are able to help us reach our highest potential. In addition to its particular energy frequency, each gemstone has a specific chemical composition, adding to its individuality, or "signature."

The Complex Signature of Gemstones

Almost every woman knows her birthstone. Actually I believe that birthstones as we know them were a marketing invention of the jewelry business. But birthstones are also an expression of our intuitive knowledge of the healing and transcending qualities of gemstones and our longing to touch the mystery of life.

There are twelve birthstones associated with the signs of the zodiac. The list of specific stones has varied throughout history, but jewelers have adopted an official list, which may not actually be based on true astrological principles. The birthstones that most of us are familiar with are based on the position of the Sun in the zodiac in the birth month. This is a good start, but it is insufficient to simply know your birth month, because your birth chart is a complex representation of many different planetary factors. In order to identify the stone that will have a distinct impact on your unique soul, it is important to study your birth chart in all its complexity. Each gemstone enfolds a specific assemblage of different planetary principles that make up its signature. The signature gives each gemstone a very different unique "personality."

Let's look at an example. The diamond, our most expensive and cherished gemstone, occurs in colorless, yellow, brown, green, blue, reddish, and black. It holds qualities of Mars, Venus, the Sun, Saturn, or Uranus, depending on its color. Diamond is the hardest of all stones and, therefore, it tops the Mohs' hardness scale[1] with a hardness of 10. This makes it a Saturn stone, because Saturn represents the principles of crystallization, hardness, and contraction. At the same time, diamond has the highest brilliancy and reflection of all gems, which clearly makes it a Sun stone. The chemical composition of diamond is crystallized carbon, which is again Saturn. The crystal structure is mainly octahedrons (a double pyramid), which makes it both Mars and Venus. Mars is reflected in the pointedness of the pyramid structure, reminiscent of a sword. The double pyramid is a very balanced form mirrored around the midline in keeping with the balance and harmony of Venus. To give an example of the relevance of color, a blue diamond is associated with Uranus because Uranus is blue.

How can we use this information to our best advantage? The signature of a gemstone tells us which principles the stone is beneficial for. For example, a Saturn-Sun conjunction in Libra or Aries points to a challenge surrounding the conflicting principles of constriction versus life force. The diamond's signature incorporates both Saturn

---

[1]. In 1822 the German mineralogist Friedrich Mohs devised the Mohs hardness scale, a simple method of rating the hardness of minerals on a scale of 1 to 10 by comparing their resistance to scratching. Gemstones used for jewelry need to have a scratch resistance of 6 or more. Diamond is the hardest mineral, with a scratch resistance of 10. Other examples of gemstones with suitable hardness for jewelry include corundum (rubies and sapphires (9); Topaz (8); Quartz (including Amethyst and Citrine (7); and Feldspar (6).

and Sun principles, making it the perfect stone to smooth out this conflict between constriction and expansion. Bringing diamond into your energy field can balance those conflicting forces within you. For example, the diamond could enhance and focus creative self-expression in your personal and professional life. It also could help you get started on new projects and align with your intention. With the diagnostic help of the signatures of gemstones, we are equipped to select exactly the right stone that is needed in each moment of our lives.

Each Stone has a Unique Personality

While each gemstone has a signature that reveals its qualities and the impact it can have on energy, in my experience there is more to it than that. I believe that not only does each type of gemstone have characteristic qualities, but that each individual specimen of a stone has its own distinct appearance and personality as well. So you must choose not only an opal, for example, you must choose the opal that is just right for you.

Tuning into your intuition will help you to find the stone that is the most balancing for you, and I usually advise choosing the stone that appeals to you the most, the one that you find most beautiful and enchanting. A gemstone's beauty is the language it uses to reach out to you and communicate with you. There are other ways to determine which exact gemstone is the most beneficial for you, but the language of beauty is the simplest one there is.

Moving through Transits with the Help of Gemstones

Armed with an understanding of astrology, you will be able to find balance during the inevitable difficult periods, which are characterized by the transits that activate particular planets in your horoscope. If understood properly, these times can become opportunities.

From the moment of our birth, represented by the birth chart, the planets continue moving through their positions in the zodiac. When a transiting planet stands at a specific angle relative to its original position in the birth chart, it will come into influence in your life, and studying these transit aspects will reveal the planetary forces in effect at a specific time. Then you can be alert for both opportunities and challenges that may arise during a particular period, and gain understanding about times when things seem particularly difficult. With a new and deeper perspective on the dynamics at work in your life at present, you are better equipped to make conscious choices and decisions. You can support those intentions by choosing a gemstone that has exactly the right signature for the moment.

Transits of the outer planets—Saturn, Uranus, Neptune, and Pluto—are often fraught with difficulty. You will want to consider using opal to ease these periods. Opal does not have a crystal structure—its molecules are arranged in layers consisting of tiny spheres of silica that create an ever-changing rainbow of reflected light. These qualities of change and movement make opals the most wonderful allies for times and processes of transformation and growth.

I hope that this article inspires you to look more deeply into your relationship to the cosmos, the course of your life, and your personality. It is possible for you to take charge of your life by harnessing the ancient wisdom of astrology and the healing energy of beautiful gemstones and crystals.

Whenever I hit an obstacle on my path, I use astrology to help me find deeper insight into the principles underlying the obstacle. Astrology helps me to stay calm; it also empowers me to action during transitions from an outdated self that no longer serves me, to a new self. Astrology is simply a means of achieving higher consciousness and sensing the divine presence in your life.

The beautiful gemstones are a gift and a tool for that inner journey of healing and balance through raising your consciousness.

I wish you a multitude of joyous experiences and many blessings with your new friends, the precious gems in your life. Enjoy your journey!

The Astrology of Social Networking Sites

by Robin Ivy

Where can you meet like-minded people from all over the world, market your music, book, or artwork; and, simultaneously get back in contact with old friends, share your newest pictures with family, and make plans for tonight without a single phone call? The answer is online. Facebook and Myspace, the most popular worldwide networking sites, are a modern phenomenon. These free, interpersonal Web sites are the most frequented hangouts on the Internet. With so many million subscribers and thousands of new users each day, it's not possible to publish an accurate figure here! Registered users can, at no cost, message each other, market their work, share photos, and preview music. The two sites differ in appearance and some features, though both provide a social network and opportunity to stay updated with friends, join groups with common interests, and send and receive invitations to events.

A Uranus in Pisces Revolution

A venture into a MySpace or Facebook profile, complete with songs, photos, and whatever personal information one desires to share, is a trip into an individual's world. The MySpace network gives space for self-expression with blogs, videos, and creative layouts to express individual tastes. Once you create a profile on either site, a collage of your interests, talents, experiences, and thoughts becomes available to anyone who wishes to "open the box." From the practical, more straightforward nature of Facebook to the open format of MySpace, each sign of the zodiac can tailor the sites to suit his or her needs.

As far as astrological timing, it's of interest that the sites debuted in 2003 and 2004, after Uranus moved direct in Pisces. The planet of technology and ruler of Aquarius, Uranus is also a higher octave of Mercury, the messenger and communicator. The role of Uranus is to shake things up, push for new methods, and expand global consciousness. In Pisces until 2010, Uranus will continue to impact music and the arts, self-expression, intimacy, fantasy, and secrets. Clearly, MySpace and Facebook are Uranus in Pisces breakthroughs, shattering the norm and changing the way we communicate and make friends! Online networkers expand their number of contacts by adding friends or family, and by inviting or accepting friends of these friends to create a larger network.

Uranus is computers, connections, and large groups, while Pisces is a personal and reflective sign. Combine these two, and you get individuals sharing feelings and experiences through their status updates, blogs, and bulletins to friends. From recalling a bad day to sharing poetry or song lyrics to honoring a loved one who passed on with video or photo tributes, Internet social sites have given us license to express and reveal in a way that was never possible before.

In the privacy of one's own home computer, a profile is created for "friends" to view, and depending on one's personal privacy settings, for anyone who may stumble on it to discover! Facebook and MySpace profiles are basically public journals and calendars made available through technology. And in the nature of Uranus, both have disrupted the status quo, created controversy, broken down communication barriers, and in some cases exacerbated the gap between generations accustomed or unaccustomed to the use of computer technology in everyday life. Let's peer further into the history and astrology of the two outlets and project how each Sun Sign matches up to the features and personalities of MySpace and Facebook sites.

MySpace.com

MySpace, incarnated in July of 2003, exhibits the networking power of the Leo/Aquarius polarity with the creative nature of Pisces, the dominant astrological forces of that summer. Mercury, Jupiter, and the Sun toured Leo, and planets Mars and Uranus resided in Pisces all month long. Neptune was well into the degrees of Aquarius, across the zodiac from Leo planets, and directly opposite the planet of communication, Mercury, midmonth. The environment was ripe for an outlet like MySpace and the astrological timing was in place. A bit more history illustrates the growth of this interpersonal megalopolis.

MySpace.com was once an online storage and file-sharing firm that failed in 2001 due to lack of funds and poor service. Prior to that, the domain was home to a Web design firm. As Uranus moved into Pisces for a seven-year stay, MySpace as we now know it began. In July 2003, MySpace.com launched as the brainchild of Tom Anderson, a musician with a film degree. This Libra cofounded the site with business partners and he remains a visible part of the site as every MySpace user's first "friend." Born October 14, 1975, Tom grew up meeting friends and chatting online, and his vision was to create something better and more efficient than the networking sites he had frequented. Another mission of the new MySpace was to give musicians, artists, and bands an active online presence for networking and for communicating with a wide audience of potential fans. His idea filled a niche and it's been wildly successful. MySpace Music now has its own division where bands, record labels, and promoters post news, information, and music for users to sample and post on their own sites. Meanwhile, as of this writing, MySpace has become the top social networking site in the world and is the fifth most popular English-language site of all.

MySpace has a strong Leo chart. Jupiter and Mercury are extroverted and creative in Leo. Check out random profiles, and you'll see that people go to great lengths to individualize their cyberspace homes in grand Leo style. Graphics, both borrowed and original, help illustrate the interests and personality of the user. From "Which Tarot Card Are You?" to "What Color is Your Heart?" you can find quizzes, surveys, celebrity photos and other ways to "pimp" your site in a variety of online locations. MySpace has few technological or content restrictions, so basically if you can find it or design it, you can post most anything on your space.

The Leo "showbiz" personality is also part of the attraction to the site. MySpace users can add musicians, authors, athletes, teams, radio and television personalities, and other famous friends who have profiles. Interaction with a "star" can range from receiving general updates to bulletins or messages about secret, friends-only events to accessing their blogs or, in some cases, maintaining contact through personal comments and messages. A MySpace profile is practically a must now in the entertainment industry, and entertainers and artists benefit from the free and extensive promotion a site provides.

Like all good things, MySpace also has a dark underbelly. The site's users are free to create their own fantasy worlds, which can include altering their ages or using false identities. For example, there are many sites where pranksters pose as famous people and seek legions of "friends" falsely. A common deception is the age change, where the young increase their age or older users opt to age down by entering false birth dates in the sign-up process. Whenever Neptune is involved, there's a sense of illusion or fantasy present, and in the case of MySpace, Mercury opposed Neptune at the time of inception. Some of this fantasy is harmless, though other more destructive forms of misrepresentation have been cited. Neptune opposite Mercury is a warning to use communication wisely to avoid scandal, gossip, or danger. MySpace users are wise to protect themselves from poseurs, predators, or known detractors who could use information with harmful intent.

Though both sites have implemented privacy settings, Facebook users seem to be a more cautious bunch. If you browse for potential friends on both sites, you will find many more on MySpace whose profiles are easily viewed.

Facebook.com

Facebook shares the Aquarian community aspect, with the Sun just a few degrees from Neptune at its launch on February 4, 2004. The site was conceived by Mark Zuckerberg, a Taurus and one-time Harvard student, who designed it as a networking space for Harvard. Too popular to stay exclusive, the site soon spread to other Ivy League and Boston campuses. By September of 2006, Facebook was available to anyone over the age of thirteen who had a valid e-mail account. Often considered a more elite or intelligent networking site, Facebook has a more practical, streamlined system designed less for creative expression and more for instant connection. A notable astrological influence on the date of Facebook inception was Mercury in Capricorn, which may account for the down-to-earth feel of the site. Mercury in Capricorn is less showbiz and more business-like and concrete. Facebook focuses on what's going on in the moment and is less about color, graphics, and personal style, reflecting a more Capricorn mentality.

Click on Facebook "Home" and you'll find a news feed feature, a basic premise of the site, which keeps friends informed of their contacts' updates, such as where they are currently, who they are now friends with, and what pages they have visited. This has sometimes been linked to cyberstalking; however, on Facebook, one can only add friends linked to work, school, or a social network, and privacy settings are available.

Subscribers can control what type of information is shared automatically and in turn prevent friends from seeing updates as desired. A Facebook profile, limited to professional or school network, with or without additional privacy settings, is less available to random browsers than one on MySpace. Still, with a strong Neptune influence, individuals need to use caution to avoid deception and distorted realities that could result from online interactions. College admissions officers and prospective employers now screen candidates' profiles as well, and many agree that it's best to post only what you're willing to be judged upon, or to keep your profile closed to the public.

Many people enjoy keeping profiles on both sites, possibly with different frequency of use and often for different intents. Our Sun sign may very well factor into which site we feel most comfortable with or which best suits our use. Style, habits, and purposes all factor in to Sun signs online.

Aries
A speedy way to communicate is ideal for Aries, who wants to get the message across and get on to other things. Aries may not create the most extensive photo album or reveal all in a blog, but she does like the instant gratification of broadcasting plans and events to a large group of friends simultaneously, followed by quick replies. Aries is a Facebook on-the-go kind of girl, with a handheld she can check at a moment's notice. The MySpace world may require too much time and effort.

Taurus
Taurus is generally a cautious user of networking sites. Slower to warm up and less likely to give out personal information, Taurus will lean toward a profile for friends only, with privacy settings in place. Building a comfort level, Taurus will then be happy to attract old friends back into his circle and develop an enjoyment of checking in, posting photos, and keeping in better touch, something he used to do mainly by telephone. Facebook's streamlined nature may suit Taurus best.

Gemini
Gemini is a sign that will happily take advantage of all online outlets. Facebook may be the social outlet used to plan with friends, find the best party, and keep in continuous touch. Myspace combined with a site like Live Journal provide the blogging space and outlet necessary for Gemini to spread videos, photographs, and news stories of interest. Once Gemini gets started, the online world can be central to his universe. Granted, it may take a while for the initial interest to develop, but once Gemini realizes there's a way to socialize while working and a method for marketing while making new friends, Gemini will wonder how we ever knew life another way.

Cancer
Reconnecting with old friends is a huge draw for Cancer, and Facebook seems to satisfy that need best. Since Facebook is organized by networks, such as alma maters and hometowns, the site can become

one big reunion online. Cancer likes that! While MySpace provides a good creative outlet for Cancer musicians, writers, or artists, Facebook has the personal side a Cancer relates to. Going through pictures of high school buddies organized in albums appeals to Cancer's innate connection to the past. A Facebook trip down Memory Lane will hook Cancer in for future visits to friends' profiles.

Leo

The wide open nature of MySpace has Leo appeal. With videos, blogs, photos, and creative layouts, Leos love the ease of access and creative expression factor. Then there's the lure of unlimited friends and contacts, connection to showbiz personalities and performers, and the ability to showcase one's own talents and tastes. Leo is in Internet heaven in this domain. As one of the zodiac's political, creative, and social beings, Leos won't hesitate to frequent more than one site.

Virgo

Virgo is a taste-tester in the online world. He'll sample networking sites, find out which fits his options best, and tend to use it minimally to communicate and plan in practical ways. Facebook may be more of a home to Virgo, who's satisfied with the more minimalist approach, appreciates the privacy of smaller networks, and doesn't need flash and glitter on his page as a rule. Virgo is one sign that will realize and appreciate the business possibilities of both MySpace and Facebook. Ruled by planet Mercury, Virgo will combine online friendships with building a clientele or marketing services or goods.

Libra

Both the founders of MySpace and Facebook were born under signs ruled by Venus, the zodiac's charmer. Like designers Tom and Mark, Libra will revel in the social aspect and also maximize use of Facebook and MySpace for both professional and personal reasons. Ruled by air, Libra is likely to use the "chat" function that allows instant conversations with online friends. From making plans for Friday night to sharing shopping or sales tips, the profile photo and instant connection has Libra appeal. Libra may be your friend who uses sites to promote music or sell from her eBay store, finding ways to innovate and combine business with pleasure.

Scorpio

Having a dedicated space for self-expression is a great Scorpio outlet. A Scorpio writer can blog or try out his new stories on friends. Scorpio musicians can post songs, and videographers and photographers can create galleries of their works. Scorpio will appreciate features of both sites, using MySpace for creative pursuits and Facebook, with privacy settings, for contact with true friends. Scorpio appreciates you not leaving lots of public messages and comments. Make plans and send messages using the Inbox instead.

Sagittarius

Sagittarius is a casual user of online networking sites. In fact, many Sagittarians may find they don't have the time or the attention span to bother. On the other hand, the social network is appealing to Sagittarius, and once hooked she will value the speed and ease of connecting with friends. Facebook mobile access on the handheld is made for Sagittarius's personal style, and with no downtime necessary, Sagittarius can communicate on the fly. Sagittarius musicians and writers are the exception, and they will use MySpace and other online outlets with more enthusiasm to promote their work.

Capricorn

Capricorn shares that Facebook quality of streamlining communication. Remember that Mercury visited Capricorn when Facebook was born. Capricorn is a sign that will use online networking for both business and pleasure and may choose to keep separate sites for each. For professional reasons, Capricorn appreciates being able to market and manage a contact list so easily. These sites are just another way to do business for a sign that loves to diversify. For friendship, Capricorn may not be out to meet new people as much as keep up with high school, college, and other friends and acquaintances. Capricorn values the past, and previous to online networking, catching up with old friends required much more work. For Capricorn, Facebook and MySpace use has to be efficient and personally satisfying to keep his attention.

Aquarius

Aquarius, the sign of friendship and community, vision, and technology, probably wonders how we ever functioned prior to online networks! Astrologically, both major networking sites have strong Aquarius charts, illustrating the Aquarian aspect of the online world in general. Facebook and MySpace groups, where like-minded individuals connect, will appeal to Aquarius, who is always interested in sharing information and expertise. From maintaining friendships to creating social calendars, broadcasting music and events, and maintaining galleries or blogs of creative work, Aquarius is all in!

Pisces

Pisces folks vary in the way they feel about online social sites. Some Pisces love another way to communicate with friends while others feel it's no substitute for a real phone conversation or face-to-face meeting. Creative Pisces revels in a place where you can share art and music with ease! Then there are the Pisces who act as voyeurs, scanning sites and profiles but participating minimally, out of fear or uncertainty. Pisces is a complex sign when it comes to online networking. Using an appropriate amount of caution is a good idea for Pisces, Neptune's sign, since they are vulnerable to any kind of fantasy world. Once comfortable, Pisces enjoys having another way to keep in touch with dear friends who are far away.

Leaders Follow the Sun
by Bruce Scofield

Astrology is based on correlations between alignments in the solar system and social and individual behaviors and processes on Earth. Each planet, and the Sun and Moon, have their designated spheres of influence on the social level. For example, the Moon correlates with the public, Mercury with writers and speakers, Venus with females and peacemakers, Mars with warriors, Jupiter with judges, Saturn with scientists, Uranus with agitators, Neptune with artists, and Pluto with corporate powers. The domain of each planet is, of course, more complex than this short list, but you should get the idea. It is the Sun, the symbol of leaders, that is the subject of this article. It is to the Sun that we look to understand moments when leadership is required, and also the effectiveness of that leadership. We will be looking at the Sun's alignments with the other planets in 2010 to better understand the subtle patterns of leadership challenges.

What do I mean by leadership? On the social level there are times when crises occur and somebody has to make a decision and initiate some kind of action. The most obvious examples are heads of state. During a major social crisis, such as a hurricane or terrorist bombing, we look to our elected leaders to do something about the problem. But sometimes leadership is more subtle, such as when leaders make choices that are not an apparent response to an immediate crisis. They may solve the problems later or cause a crisis in the future. For example, the signing into law of a bill that changes things—not right away, but eventually—is something that political leaders do frequently, and sometimes their judgment is not so good. Leadership also happens in other social situations. Every state has its governor, every town its mayor, every school its principal, every business its boss, and every team its captain. All people in leadership roles are affected by the alignments of the Sun, Earth, and the other planets.

As the Earth orbits the Sun each year, alignments with the other planets form and dissolve, like doors opening and closing. But it is never that simple. There are times when a solar alignment with one planet is increasing in intensity while an alignment with another is fading out. Sometimes two or more alignments occur at the same time, and sometimes there are no alignments for several weeks. The most important alignments, or aspects, are the conjunction, opposition, trine, and square. The opposition is particularly powerful as this is when the Earth is sandwiched between the Sun and another planet.

Such alignments occur only once a year, and not at the same time. For example, each year the Sun will stand on one side of the Earth while Saturn will stand on the other. This is the opposition of the Sun and Saturn (from our geocentric point of view) and the date of this opposition in the calendar moves ahead by about two weeks each year. The Sun and Earth also align with Jupiter, Uranus, Neptune, and Pluto each year, as well. These dates are not consistent with the seasons and must be calculated in advance. Solar alignments with the inner planets, Mercury and Venus, are limited to the conjunction and, in the case of Venus, the semi-square aspect of 45 degrees.

The most frequent solar alignment is with Mercury. It is when Mercury is aligned with another planet that leaders make important announcements; it is a time of major communications, diplomatic activities, and decisions. Such things may make the news and leaders will often speak with great authority. Very often, transportation matters become an issue and may get special attention.

There are two kinds of solar conjunctions with Mercury and they have very different flavors. The superior conjunction (when the superior body—the Sun—is between the Earth and Mercury) occurs when Mercury is traveling direct in the zodiac. In 2010, superior conjunctions will occur on March 14, June 28, and October 17. Around those dates, give or take a few days, expect to hear judgments and pronouncements that come after some deliberation.

When Mercury is retrograde and forms a conjunction with the Sun it is called the inferior conjunction because the smaller body, Mercury, comes between the Earth and Sun. At these times, January 4, April 28, September 3, and December 19, leaders will be stating their thoughts, but with a bit more intensity. Misunderstandings between leaders and botched plans can occur within a few days of this conjunction, but it is also possible that some bold new initiatives may be announced. The inferior conjunction is a time of risk.

Tracking Our Leaders

We'll now track the passage of the Earth and Sun relative to the other planets for 2010 and describe what kinds of issues leaders may face. Give a few days either side of the dates for the astrological effects to be noticed. It may be that nearly all leaders will feel these aspects in their realms, but only a few on the world stage will make the news. Also, keep in mind that these are not the only days in which leaders act. There are other aspects that I have not included, but if an important event occurs near one of these dates, the description will apply. As we pass through the year we can all judge how well our leaders (and perhaps you, especially if you are a Leo, or have Leo rising) have responded.

January 11: Sun Conjunct Venus
This is the superior conjunction with Venus, when the Sun stands between Venus and the Earth. Leaders may have real problems now in coming to decisions, or the decisions will have been very difficult and concessions and compromises may be needed. There may even

be an ethical or moral quality to choices or announcements being made now, but it will most likely be a time when conservative elements dominate over more liberal options.

January 24: Sun Trine Saturn
Leaders will be engaged in serious discussions and will tend to make very realistic and practical choices. It is a time of solutions to scientific problems and also issues pertaining to standards of all kinds. Boundaries are reinforced and old issues are revisited and put on the right track.

January 29: Sun Opposite Mars
This is a time when courageous leadership is demonstrated or talked about or, conversely, it may be a time of conflict and error. Leaders will need to define their territory and, if challenged, they will fight. This theme will be reflected in sporting events, military actions, and political debates. Decisions will be made quickly and efficiently.

February 14: Sun Conjunct Neptune
This can be a challenging time for leaders without a vision, but a wonderful time for those with one. Idealism prevails, but this can bring with it episodes of misjudgment, false accusations and spiritual or religious misunderstandings. Leaders may give more attention to the arts and religion now, perhaps meeting with leaders in those fields.

February 28: Sun Conjunct Jupiter
Expect leaders to exaggerate things now and take dramatic actions, even though they may not be especially important. It's all about size and show, and leaders may use this time to advance their reputations. Many generous proposals may be offered, big initiatives launched and long trips made. Legal or moral values will probably be expressed at this time as well, as leaders seek to justify their actions by appealing to a higher authority.

March 17: Sun Conjunct Uranus
Leadership is less stable under this aspect. Some sudden moves and radical proposals may make news and shake people up. It's a time for changes, but there's always resistance to change so it behooves

leaders to be patient and move slowly if attempting to reform existing patterns. This time favors innovation, however, and a lot of progress could be made in a very short time.

March 21: Sun Trine Mars

Exemplary leadership is demonstrated or talked about and leaders state their territorial interests clearly. Sporting events around this time will be exciting and so will any political debates. It can be a very constructive time and decisions will be made quickly and efficiently. However, this aspect is concurrent with another involving Saturn, which suggests a more complicated situation.

March 21: Sun Opposition Saturn

Often leaders take a loss under this aspect, but some will benefit, if they are honest and have paid their dues. False leadership will be flushed out and defeated, and contenders may replace current champions. Countries may lose a leader in any number of ways and other leaders may find themselves having to make very realistic choices. Traditions and authorities from the past may be part of this process. The best leaders will use the past to build a future that is planned carefully.

March 25: Sun Square Pluto

This aspect intensifies the trends of the week and brings up hidden matters, secrecy, and the need for an investigation. Leaders will have to deal with unpleasant situations and make crucial choices, and they should not do this unilaterally. What is required now is cooperation, but with the other aspects also in operation, there could be a breakdown of relations and conditions may be fragile. The best leaders will do what is necessary but only after sharing their concerns with others and coming to a group decision.

April 25: Sun Trine Pluto

If leaders move to enhance their power (or that of the group they represent) at this time, they will probably succeed. They will use probes, investigations, cleanups, and all kinds of renewals to create change in their world. It all comes down to intelligent strategies and consistent execution, but that is what it takes to move things along and for stagnant situations to evolve.

May 4: Sun Square Mars
Leaders may pick fights, or be drawn into them; it may be a time of quarrels, conflicts, strife and disruption. Courageous leadership may be demonstrated or talked about, but more likely there will be a need to respond strongly to events like accidents, disasters, or political tensions. This theme will also be seen in sporting events and political debates that occur near this date. Decisions will be made impulsively and urgently.

May 18: Sun Trine Saturn
Leaders, who no doubt have been thinking about how to make things more secure for some time, will now be engaged in serious discussions and will tend to make very realistic, conservative, and practical choices. It is a time of solutions to practical and scientific problems and also to issues pertaining to standards of all kinds. Boundaries are reinforced and old issues are revisited and put on the right track.

May 19: Sun Square Neptune
This alignment will challenge leaders in strange and subtle ways. It may appear on the surface that nothing is happening, but on deeper levels leaders may be weakened, harmed, or inhibited in some way. Pressures from religious or political extremists will rise and their needs will have to be considered. Much of the future is uncertain now and only the best leaders will navigate things successfully. More likely, this can be a time of misjudgments, false accusations, and general confusion.

June 19: Sun Square Saturn
Setbacks and roadblocks will confront leaders at this time and they will need to be very focused and disciplined to bring the best out of the existing conditions. Leaders will be held responsible. Fear is prevalent and delays occur. The best leaders will not do anything radical, and stay within the normal boundaries when making hard decisions. This could be a time of many changes as this alignment begins a series of major solar aspects over the next few days, which will make things very complex.

June 19: Sun Trine Neptune

This alignment favors leaders who have strong beliefs or artistic inclinations. They will lead through compassion and sensitivity, not force and domination. There is a striving for ideals at this time, but this is complicated by the other aspects in operation.

June 21: Sun Square Uranus

Leaders will be struggling with surprises and unexpected events, and this will test their abilities to make choices and decisions. An atmosphere of uncertainty will surely call for some action, but the best leaders will do this in a temporary way and allow conditions to change before they commit themselves to a permanent plan for change.

June 23: Sun Square Jupiter

Now circumstances are highlighted by moral or legal issues and leaders will need to take these into consideration and act in ways that will not come back to haunt them. Doing too much at this time may prove unhelpful, but something has to be done or situations will grow too large in certain ways.

June 25: Sun Opposite Pluto

This aspect brings the trends of the past week to some kind of climax. Leaders will have to make crucial choices and they will need to do this in concert with others. It could be an opportunity for high level cooperation, or a breakdown of relations. Conditions may be fragile. The best leaders will do what is necessary but only after sharing their concerns with others and coming to a group decision.

July 23: Sun Trine Uranus

Leaders will now focus on new or radical ideas and solutions to problems. Technology may play a role and the overall effects of actions taken now will be liberating. Those with vision and foresight will be acknowledged and their ideas put into practice.

July 26: Sun Trine Jupiter

Expect leaders to express confidence and use this time to advance their reputations. They may show great generosity and long-range vision, and big initiatives may be launched and long journeys made. Legal or moral values will probably be expressed at this time as well, as leaders seek to justify their actions by appealing to a higher authority.

August 26: Sun Trine Pluto

Some leaders may make a bid for power at this time, and they will probably succeed. The use of probes and investigations will support cleanups and all kinds of renewal to create change in the world. It will be intelligent strategies and consistent execution that moves things along and allow stagnant situations to evolve.

September 21: Sun Opposite Jupiter

Circumstances may be highlighted by moral or legal issues and leaders will need to take these into consideration and act in ways that will not come back to haunt them. This is a time of profound decisions that must take into account many factors, and this will not please everyone. Doing too much at this time may prove unhelpful, but something has to be done or situations will become lopsided and expansive.

September 25: Sun Square Pluto

Leaders may need to handle situations that have not been on the radar screen and they may need an investigation to do this. They will need to deal with unpleasant situations and make crucial choices and cooperate as much as possible. The best leaders will do what is necessary but only after sharing their concerns with others and coming to a group decision.

September 30: Sun Conjunct Saturn

Responsibility is the theme now and the past leans heavily on the present. The flavor of this time is conservative and fears abound. The best leaders will harness this leaning and give it some structure and form. They should see solutions to practical and scientific problems and also to issues pertaining to standards of all kinds. In general though, boundaries are reinforced and old issues are revisited and hopefully put on the right track.

October 19: Sun Trine Neptune

Leaders who have strong beliefs or artistic inclinations will shine during this alignment. They will lead through compassion and sensitivity—not force and domination—will strive to realize ideals. They might display a bohemian side and make unrealistic decisions.

October 28: Sun Conjunct Venus

This is the inferior conjunction of the Sun and Venus, and it signals a time when leadership can be impulsive and prone to errors in judgment. Mistakes could be made and leaders may have to atone for their sins. The people of ancient Mexico and Central America regarded this conjunction as very dangerous and they believed that it "struck down" people in high places. I have found this to be true in many cases where leaders were discredited, removed from power, committed crimes of a moral nature, or made other errors of judgment.

November 15: Sun Trine Jupiter

Leaders will express confidence and use this time to advance their reputations. Displays of generosity, statements of long-range vision, embarking on long journeys, and big initiatives may be launched. Legal or moral values could be expressed at this time as well, as leaders seek to justify their actions by appealing to a higher authority.

November 18: Sun Square Neptune

It may appear on the surface that things are fine and that nothing is happening, but on deeper levels leaders' influence may be weakened, harmed, or inhibited in some way. Pressure from religious or political extremists will rise and their needs will have to be considered. Much of the future is uncertain now and only the best leaders will navigate things successfully. More likely, this can be a time of misjudgments, false accusations, and general confusion.

November 19: Sun Trine Uranus

The focus is now on new or radical ideas and solutions to problems. Leaders will need to deal with technology, they may initiate idealistic programs, and they may strive to free things up in one way or another. Those with vision and foresight will be acknowledged and their ideas put into practice.

December 16: Sun Square Jupiter
Leaders will need to confront moral or legal issues and act in ways that will not come back to haunt them. Doing too much at this time may prove unhelpful, but something has to be done or situations will grow too large in certain ways. Certain risks may not pay off.

December 18: Sun Square Uranus
Leaders should expect surprises. Along with the above aspect, leaders will be struggling with unexpected events, and this will test their abilities to make choices and decisions. An atmosphere of uncertainty will surely call for some action, but the best leaders will do this in a temporary way and allow conditions to change before they commit themselves to a permanent plan for change.

December 26: Sun Conjunct Pluto
A lot could be happening now that puts pressure on leaders, but not a lot of it will be visible. In some ways this aspect marks a change of substance in leadership at all levels. Leaders will have to make crucial choices and they will need to do this in concert with others. It could an opportunity for high-level cooperation, or a breakdown of relations—conditions may be that fragile. The best leaders will do what is necessary but only after sharing their concerns with others and coming to a group decision.

Home Makeovers on a Streamlined Budget

by Alice DeVille

The year 2010 dawns. You clink the champagne glasses together in anticipation of a new cycle, and your thoughts turn toward cherished plans for the year ahead. Images float by on your personal wheel of life. Lo and behold your residence comes into view! That amazing vision looks like the new home office you want to create. Next come the "snapshots" of disorganized closets, stark white living room walls, and outdated kitchen countertops. Wondering if your checkbook can bear the expense? Looks like you're going to be setting some priorities. Now is the time to reflect on last year's household spending and focus on current plans for improving your home.

Does a resolution to redecorate top your list? If so, are you aware that astrology highlights current planetary trends that are likely to affect your purchasing power? The year starts off with the planet Saturn in harmony-loving Libra, a Venus-ruled sign. This combination of Saturn (the natural ruler of reserved Capricorn) in indulgent Libra has very different properties. Saturn likes to remind you of where you have limitations, while Libra enjoys all aspects of manifesting beauty and savors every frill. So regardless of your Sun sign, your first step is to take inventory of your budget.

Agonizing over what change will increase the value of your home? When it comes to your home, every new purchase can become a gut-wrenching investment decision under the influence of Saturn in Libra. Saturn says: "Be practical and make a choice." Libra says: "I want it all." Can you feel the conflict building? This article takes the pain out of home makeovers and offers tips for resourceful renovating.

Adding Glamour on a Budget

Dare to be Different
Unusual touches give a room or section of a room character. The smallest nook comes alive with the right decorator touch. Pillows with ornate beading energize rooms; so does a patterned carpet. Wallpaper makes a statement in the powder room or as a border in bedrooms.

Raid the attic or visit a flea market to find new pieces once you decide on a theme. Rooms or entrances with many angles and doors call for careful balance in texture and color. Give a small entry foyer that opens to a living room privacy and depth by adding sweeping, fringed drapes fastened gracefully to create the "entrance." Set the stage by painting the foyer a deep color and cover a solid wall with three columns of coordinated framed art. Add a row of larger framed art above the columns using two pieces. Above the entry door, place a piece of metal artwork or rectangular-framed piece flanked by sconces. The result? It will look like you added a room without hiring a carpenter.

Add Lighting Enhancements
Focus on room lighting and create a dramatic new look. If lighting is too dim, the room lacks energy. Even with the most beautiful furnishings, a dark room lacks warmth and discourages social interaction. Any number of reasons contribute to this ho-hum look: the wattage in lamps is too low, shades are too dark, position or style of

lamps is not conducive to reading, rooms have ineffective lighting, most of the floor lamps throw light to the ceiling (or conversely, no lamps throw light toward the ceiling), lamp choices match color scheme but lack utility, lamps overwhelm tables, the chandelier is either too small or too large or it does not fit the décor, light bulbs are dusty or dull film covers chandelier crystals, kitchen lighting is inadequate for the workspace, bathroom lighting either "cooks" you with too much heat or it is so dim you can't see well enough to pluck your eyebrows.

Go for the Glow
Lighting deficiencies can easily be corrected most of time. Some can be done on a small budget if you tackle them gradually. First clean every light fixture with a microfiber cloth, glass-cleaning wipes, dampened dust cloths, or a long-handled duster. Begin by taking inventory of your family or living room. Reposition lamps to enhance lighting, replace dark shades with lighter, unpleated shades (pleats collect dust). Look for lamp sales, visit thrift shops or go to estate auctions to find new lamps or replacements for lamps that don't work well in your spaces. For a few thousand dollars your electrician can install overhead lighting, under cabinet lighting, spot lighting, pendant lights, and recessed lighting to give your kitchen new life. Who wouldn't want to cook a meal with tailored kitchen task lighting that makes recipe testing a pleasure! Replace glaring bathroom lighting with three- or four-light sconces, recessed ceiling lights, or soft tube lighting to give a modern esthetic look to your bath.

Paint Can Totally Change a Room
Not just any paint will do, though. The right shade of paint that was designed to make a statement and complement your walls can dramatically change the feel of a room. Sometimes just painting one or two walls is all you need. A current trend is to use eggshell finishes because they are washable and look new longer. This finish is fine as long as the walls are fairly smooth; if they are not, however, every imperfection shows up. Use flat paint instead. If the trim could use a makeover, semigloss provides nice accent sheen and is easy to clean. A satin finish has a quiet shimmer that can be washed often. It works well in kitchens, bathrooms, and homes with small

children. If you are tired of white or off-white trim, try a striking accent color. Outline the crown molding and chair rail. Natural wood tones look striking in many rooms. You'll receive many compliments for your innovation.

Buy the best paint your budget can afford. There are higher quality paint grades, even for the budget-minded, that are easy to work with and attractive.

Install New Cabinet Handles
Swap out old, broken, or tarnished handles with handsome new replacements. That way you don't have to go through the expense of installing new cabinets. It's easy, eye-catching, and affordable. In most instances, you can replace the entire kitchen set with good quality pieces for under $350. Be sure to take measurements so you don't have to fill in extra holes or refinish sections of cabinets that used larger hardware.

Pressure-wash Your Home and Add Exterior Enhancements
Your home may look dingy and you think it needs a new coat of paint. Using a good power washer will save you a considerable amount of money and often has your paint job looking as good as new. (If you have a lot of peeling paint or damaged wood, this idea won't work.)

Any one of these will dress up your exterior: a nice new mailbox, stylish address numbers, or colorful potted plants near your front door. Each is an inexpensive way to give your home a welcoming feel.

Wash Walls and Windows
Just like your exterior paint job, the interior gets dingy. You can often return the luster of old paint with a little elbow grease. Keep Magic Eraser® on hand for quick touch-ups. Crystal-clear windows make rooms brighter and improve the look of your home's exterior. Window washing firms offer seasonal sales, or do it yourself using ammonia and water for a streak-free shine.

Maintain Your Yard
Avoid an unkempt façade. Nothing leaves the impression of a home in disrepair more than a poorly maintained yard. Haul away trash and debris, weed planted beds, and prune overgrown shrubs

and trees. Add bright flowers and dust off the doormat to give your home a welcoming feel.

Paint or Replace the Front Door
If the front door faces the sun, color fades. A new coat of paint adds brightness and stimulates the prosperity-generating aspect of your entryway, a principal rule of feng shui. Tarnished hardware spells neglect. Opt for a lifetime finish on the replacement hardware and you'll never have to worry about polishing brass. Warped doors are an energy drain and no amount of painting is going to remedy the look of an ill-fitting door or damaged frame. Shop around for a good deal on an insulated replacement in fiberglass, steel, or wood. Include a leaded glass insert for a decorator touch. Order the hardware package that suits your needs, including double-security lock sets, brass kick plates, door knocker, and viewfinder. Most manufacturers customize your order before installation to minimize messy on-site adjustments.

Easy Organizing

Although this topic deserves an entire article, I've included a few organizing shortcuts to make your life easier. Purchases are minimal, but the peace of mind that comes with good storage solutions is significant.

Nothing is more frustrating than not being able to find the pots, pans, lids, and serving pieces when you're planning a big dinner party. So store pots and pans near the stove and oven for easy access; then, store cookware according to function. Place colanders with pasta pots, mixing bowls with bakeware, flat items like cookie sheets and cutting boards in easy-access narrow cabinets where you can stand them upright. Toss lids with loose or broken handles and warped pots and keep only the cookware you really use. Donate the rest to charity.

Invest in a roll of rubberized shelf liners to protect pots, pans, and serving pieces from damage and noise pollution. Buy spring-loaded drawer dividers (about $20 for two) and place them in utensil drawers to separate spoons, spatulas, measuring spoons, and tongs. Eliminate lid searches by mounting a lid rack (an $8 investment at the hardware store) behind the cabinet door where you store pots and pans. Use a similarly priced tiered rack to stack frying and sauté pans inside the cabinet with the handles in an easy-to-grab position. Don't nest frequently used cookware. You'll waste time searching for the right size and wind up scratching and denting pots. If space is an issue, place paper plates between items to protect them from damage.

Smart Storage Solutions for Children's Rooms
Purchase a headboard with compartments that hold reading material, small toys, trophies, and treasures at easy-access angles. Some models even have small drawer cubbies and room to attach a reading light. Invest in a hamper that is large enough to hold the dirty laundry (round shapes in wicker work well). Buy two if you want one to hide toys.

To keep magazines and notebooks from littering the floor, attach a hanging fabric rack to a wall near your child's desk. Mount a good sized corkboard behind the desk so your child can post favorite articles, magazine clippings, photos, and memorabilia. Create a drop zone with hooks installed at eye level for a place to hang backpacks and jackets; mount an over-the-door see-through shoe holder to store items your child "grabs" before leaving for school, such as sunglasses, cell phones, gloves, or iPods. These inexpensive purchases come in color-coordinated materials; the biggest benefit is in eliminating the morning scramble and keeping the room neat.

Accentuate the Positive

Pay attention to current trends when planning your home makeover. Tailor projects to fit your unique style and financial reality. The planetary forecast exudes tempered optimism; it's time to express your dreams in fashionable, streamlined style.

Sunspot Explosion in 2010

by Kaye Shinker

*What has been will be again,
What has been done will be done again,
There is nothing new under the Sun.*
Ecclesiastics 1:9

The sunspot on April 2, 2001, was so large it could be seen by casual observers at sunrise or sunset. It frothed and fumed and danced across the Sun's equator, creating the most intense flare scientists have ever observed. The x-ray intensity was rated X22. The storm raged for six weeks and the direct result was brilliant dancing auroras that could be seen as far south as El Paso, Texas. Radio communications were distorted, satellites suffered irreparable damage, and several electrical transformers were killed.

On October 23, 2003, three giant sunspots unleashed eleven X-class flares in only fourteen days. Auroras appeared all over the world including Florida, Texas, Australia, and many places where they are seldom seen. These northern lights were almost a dark red as they spread southward across the unsuspecting populations south of latitude 45° north.

The appearance of an aurora is not dependent on solar flares. Magnetic storms occur every three or four days, igniting the auroras. Most of them dance for only a few minutes and the dominant color is green. They quickly light up the northern sky with streams of light. Great arcs of light form into dancing beams of yellow, white, green, red, and violet, only to disappear as quickly as they appeared. Sunspots, however, do create solar flares. The coronal mass ejections from these solar flares create colorful auroras that are long-lasting, intense, and brilliant.

The Earth captures and stores energy radiated by the solar wind, and it is theorized the tiny protons in the solar wind are captured and eventually form clouds that release their moisture as fine rain or mist. High sunspot activity therefore encourages precipitation, and precipitation encourages low shrubs, blueberries, and willow leaves to grow quickly and abundantly, providing food for caribou and other animals to eat. As the food supply increases the survival rate of the fawns increases.

Scientists have made various studies of the amount of sunspots and the price of various agricultural products. William Herschel in 1801 noted an apparent connection between solar activity and the price of wheat. During 2008 there were very few sunspots and the price of wheat, corn, and soybeans was extremely high. Between 2000 and 2003 these commodities were abundant and inexpensive.

Myths of the Aurora Borealis

The aurora borealis has been lighting up the skies forever, and stories abound in the folklore of people who regularly experience these northern lights. Some groups believed that the colorful displays foretell abundance and good fortune while other groups found them frightening harbingers of wars and plagues.

Northern European Myths

Scandinavian fishermen welcomed the northern lights, believing that they were the reflection of great shoals of herring and foretold large catches of fish. The Scandinavian name for the northern lights translates to "herring flash," suggesting that the fishermen believed that lights in the sky were reflections of these vast swarms of silvery herring. Norwegian sailors believed the lights were the souls of maidens, waving and dancing; and the Danes believed the lights were the reflection of the ice on the wings of swans that had flown too far north. Swedish farmers thought that the aurora meant a bountiful harvest with plenty of seeds for the following year. Finnish myths tell of foxes with tails on fire that light the northern skies. Norse mythology tells of virgins mounted on horseback and armed with helmets and spears. Their armor sheds the strange flickering light that flashes across the northern skies.

Eastern European Myths
Russian folklore told of a fire dragon that seduced women when their husbands were absent, and the Sami people felt that mocking the lights or singing about them was extremely dangerous; even life-threatening.

The Latvian myths foretold an ominous disaster when the lights during the winter were red. They believed the red was the fighting souls of dead warriors. Medieval Europe saw the lights as an omen of war, disaster, and plagues. In Scotland, the lights were known as "the Mirrie dancers."

Canadian and Arctic-region Myths
The Caribou Inuits and the Inuits of the Copper River believed that the brightest of the auroras brought good hunting and prosperity to their people. Experience taught them that caribou were abundant with bright auras and that there would be plenty of meat to feed their families through the winter. They believed that the aurora spirits were responsible for good weather and bountiful hunting.

The Canadian Inuits believed in the aurora's curative powers. They believed their shamans could make astral journeys to the aurora when it appeared. The shaman received advice from the lights and returned with remedies to treat the sick. Sometimes the shaman even received the power to rescue souls from death.

The Algonquin believed the lights were their ancestors dancing around a ceremonial fire. They told stories about the manabai'wok (giants), who lived in the direction of the North Wind. The stories explained that the manabai'wok are still our friends, but we do not see them anymore. They were great hunters and fishermen, and we know whenever the manabai'wok are out with their torches to spear fish, because the sky is bright with color over the place where they are fishing.

There are tales of the dancers fighting with each other and when the morning dew produced red lichen there had been a serious battle during the night. Even the Klondike gold rush prospectors believed that the lights were a reflection of the gold they sought.

Asian Myths
In China and Japan the appearance of the auroras foretold fertility and predicted the birth of children. The lights could also ease the pain of childbirth.

The Astrology of Sunspots

Tracing the history of high sunspot activity and planetary motion reveals an interesting pattern. When Saturn, Uranus, Neptune, and Jupiter are within 30 degrees of each other by conjunction or opposition, they seem to exert enough magnetic pull on the Sun to cause the eruption of sunspots. In the year 2000, sunspots reached a maximum and the process continued for several years as Jupiter moved away from its conjunction to Saturn and into an opposition with planets Uranus and Neptune, which were within 20 degrees of each other.

In 2010 a similar pattern will be traced by the planets. Neptune and Uranus surround Jupiter and they are 30 degrees apart, indicating high potential for sunspot activity. Throughout the year, they will be opposed by Saturn and Mars. It is a very volatile time for the Sun.

It has only been nine years since the last high sunspot activity period and not the eleven-year cycle theorized by astronomers. Of course the eleven-year calculation is an average of the known solar maximums recorded by Europeans, and seems to correspond with the almost twelve-year orbit of Jupiter. Chinese astronomers have much older historic records, and when these observations are added to the European records, a more accurate timetable for the cycle may be revealed.

Scientific Research

Until the 1980s, the auroras were a treat, like a visible eclipse or a colorful sunset. Once in a while, the ham radio operators and some airline pilots complained about the static. But as the 1980s progressed, folks began shooting electronic machines into space. Soon every magnetic storm from space became a problem and research was required.

NASA Exposes Another Mystery behind the Northern Lights
Dateline July 24, 2008

The solar storms explain why the lights occur, but what makes them dance? On February 26, 2008, NASA captured an isolated sub-storm in space. Ground-based observers noted an intense brightening of an aurora while five satellites observed the magnetic reconnection that triggers the rapid brightening and rapid expansion of the aurora toward the poles. There were twenty ground-based observers set up and a series of five small satellites to measure the northern lights and find out what was responsible for the sudden bursts of light and the dancing movement of the phenomena. "The culprit turns out to be a magnetic reconnection, a common process that occurs throughout the universe when stressed magnetic field lines suddenly snap to a new shape like a rubber band that's been stretched too far."

Solar Super-storms to Come

Scientists theorize that some of the blackouts early in this new century were the result of solar storms. They explain that the grid should have been rebuilt to avoid future electrical catastrophes. Unfortunately, there are not enough areas where electrical lines are underground. Buried wires and cables have fewer problems

Sunspots in this new century can leave quite a mess for all of us to clean up. The fall of 2009 promises to be a time of increasing sunspot activity and the resulting coronal mass ejections. With the arrival of 2010 we will begin to experience radio interference and satellite blackouts.

Recently the International Space Station designated a safe room for astronauts, anticipating that the cosmic rays from the solar storms could injure or even kill these very valuable (and vulnerable) individuals. The space station astronauts may hide in their safe room during intense storms.

The sunspot cycle is predicted to peak between 2009–2011. Each of those years, between 100 and 300 sunspots will actively transverse the Sun and send out solar flares, and some will hit Earth's atmosphere. This high sunspot activity will have a detrimental effect on the thousands of communication satellites positioned above the Earth. Engineers can turn most of these satellites to avoid a direct hit, nevertheless, a few will be hit. We can expect that cell phones, televisions, GPSs, and commercial and military satellites will suffer temporary outages. Governments as well as private corporations will employ astronomers and keep them working day and night to watch each spot and flare and calculate its projected trajectory.

One interesting new problem for this sunspot cycle No.24 will be the proliferation of cell phones. Land-line phones have underground cables and most will probably operate easily on their local networks. Though not dependent on the grid, cell phones are completely dependent on satellites. None of us will be able to get mad at our friends for suddenly hanging up the cell phone.

References

Carlowicz, Michael J. and Ramon E. Lopez. *Storms from the Sun*. Washington D.C.: Joseph Henry Press, 2003.

http://www.spaceweather.com

http://www.nasa.gov/mission_pages/themis/auroras/themis_power.html

http://sidc.oma.be/html/wnosuf.html

http://side.oma.be/html/wolfaml.html

http://johncoven.wordpress.com/2008/03/17/aurora-borealis-and-myths

Sun Sign's Orientation to Change

by Lesley Francis

Okay, for the sake of an open mind, try to forget everything you've ever thought about change. Accept one thing: it really does have a purpose, which is to get us to grow. Not just to upset our carefully laid plans or make us feel like we are perpetually living on a skateboard. The bottom line is change is inevitable. It's gift is to get you to step outside what you know and push yourself to new experiences, provided you want to do more than just get through the curve ball the Universe has just thrown at you.

Now, you belong to one of twelve Sun signs, and each one of you approaches change differently. Some view it as opportunity, others view it as trouble, still others try to avoid it. Of course, I won't name names. That would be unfair.

For each of the twelve Sun signs, change is meant to get them out of the rut that is built into their temperament. Trust me, every Sun sign has inner attitudes and beliefs they like to hang onto that leave them less able to express the best parts of themselves. So let's take a look at each Sun sign's attitude toward change, how that can inhibit their growth and, finally, the key to maximizing the best qualities of what it is to be everything from an Aries to a Pisces.

Aries

Okay, so your definition of change means activity—movement. But constantly running in circles, no matter how large they are, doesn't really alter anything. Activity might make you feel good in the moment, but when you come to a halt everything is pretty much where you left it. What you need is a goal that motivates you to extend yourself, to challenge yourself; and that, in turn, forces you to trust your instincts in the pursuit of growth. And those instincts are your greatest asset. You have a built-in capacity for taking the right risk at the right time. It's what leadership is really all about— forging into unknown territory and trusting completely in yourself. Stop worrying about whether you'll have to abandon your latest impulse. Look past the temporary exhilaration it gives you and take action that will lead to something that will profoundly move your life. The truth is you want to be like Neil Armstrong and not the guy who took a bungee jump. Summon all that inner fearlessness and show the world the path to real change. For Aries, *the key to using change to create growth is purpose.*

Taurus

Now stop pretending you're hard of hearing, or that you can't spell. I know the word *change* strikes fear into your heart. Why? Because your assumption is that change will destroy everything you've worked so hard to build. It's upheaval, pure and simple. And no self-respecting Taurus is going to put out the welcome mat for that. But, hang on a minute. Change really isn't any different than winnowing out the plants in a garden that aren't flourishing. It gives everything else a better chance to grow, to thrive. And the truth is you really love to grow things. Just remember to grow yourself as lovingly as you do everything else and you won't find change such a difficult thing to embrace or act on. In fact, you might even find yourself saying "bring it on." Okay, maybe that's pushing it. But change offers you the opportunity to finally realize that you are the rock on which you can build your life, that you are your own security and nothing can change that, only you. For Taurus, *the key to using change to create growth is elimination.*

Gemini

You really don't stop to think much about change. You're too busy gallivanting around, bouncing from thing to thing, drinking in all that marvelous experience. However, curiosity is not a substitute for change. The one thing you have going for you is that you really are built for change. You just don't always know it. So use that considerable talent for mental expansion (for more than playing Trivial Pursuit® or being mesmerized by the latest gadget) to get your motor running. In truth, it's not that difficult to get your attention. The challenge for you is to assess what's being offered and then to ask yourself whether it will help you grow or if it's just another dose of mental candy. There's so much personal growth to be mined from your experiences. Take time to put them in perspective. Get past the initial that-was-interesting response and find a context for what you've done. It will lead to a deeper awareness and a deeper satisfaction, both with life and with yourself. For Gemini, *the key to using change to create growth is reflection.*

Cancer

I know you want to barricade the doors and windows of your psyche whenever change comes a-knocking. It disturbs your inner equilibrium. Why? Because change means loss to you, and that leads to an emotional earthquake that leaves you broken. But take this into consideration. You love to nurture people, projects, and all kinds of stuff. And opening up to change can allow you to do more of that, not less. How? you might wonder. By letting go. By clearing all the emotional deadwood out of your life so you can nurture (another word for grow) in a deeper, more satisfying way. Letting go leads, dare I say it, to happiness as opposed to familiarity. And you know how you love to cling to old patterns and old experiences. You make the mistake of believing they are what's meaningful in your life when, often, they are just burdens, pure and simple. Clinging to the familiar isn't any different than rearranging the deck chairs on the Titanic. Things might look different, but the ship is still going down. For Cancer, *the key to using change to create growth is release.*

Leo

Your mantra? If it ain't broke, why fix it? You might be heard to mutter "what do I need to change?" Well, no one is asking you to change your essential wonderfulness but you might need to recognize that potential isn't the same as manifestation. The acorn is not the oak. Take a look at the fact that not everything you say and do springs from the best part of you. You have your less-than-stellar moments just like everyone else. Your challenge is you ignore them. Just like you ignore anything that doesn't support your view of the world. It's all good, right? Why? Because I am good. In fact, I am great. The need to maintain that belief at all costs leaves you at a disadvantage when change comes, because, of course, it will come. You will fight it because you believe it is telling you that something is fundamentally wrong with you. Not so. The things you use to resist change—integrity and honor, loyalty and steadfastness—are the very things that will help you grow your life. You will become the oak. For Leo, *the key to using change to create growth is humility.*

Virgo

For you, change equals improvement. And in your world that's always a good thing. Why else would you change something if it wasn't going to lead to something better? However, you always get side-tracked by two things. One, you're never quite certain you're going to make the right decision, so your commitment to change can get high-centered while you analyze, analyze, analyze. It's hard for you to accept that you cannot anticipate everything. The need for perfection meets the reality that the outcome can never really be predicted. At best, every decision is a really good guess. Second, change isn't always tidy or straightforward. It can be equal parts messy and confusing. And that just drives you crazy. You hyperventilate and become hypervigilant. Every little thing that happens has to be a sign of success or failure. You're never sure which. This makes it hard to grow because the growth of a human is not something easily measured or quantified. The solution? Relax. Embrace your capacity to make the best of any situation. It's what helps you grow away from criticism toward healthy self-respect. For Virgo, *the key to using change to create growth is adaptability.*

Libra

In your world, change is just another word for compromise, or is that accommodate, or maybe it's appease. Now, there's a trifecta and one that's certain to please everyone but you. That's okay as long as the boat doesn't get rocked and you don't have to deal with things head-on. You're always afraid that the purpose of change is to shine a light on something in your life that's not very pretty. And that's probably true. Still, you're not having any of that. So, you continue to gloss things over or you look for the tiniest of positive things to reassure you that everything is okay. And, when things keep on cracking or collapsing or becoming downright painful, you wonder why you're being picked on. However, always being nice can make you brittle. The winds of change really want to free you from this self-imposed exile and grow you into the idea that subservience does not make for harmony or balance. Take all those things you've learned about making others happy and plunk yourself in the middle. Make you the focus of your grace. See how good it feels. For Libra, *the key to using change to create growth is self-awareness.*

Scorpio

Yeah, you love change. In fact, you'll tell anybody who'll listen you're not afraid of it. As long as it doesn't expose you too much. But that's the trouble with change; it tends to blow things wide open, mangling both your carefully constructed outer cynicism and all that camouflage you invest so much emotional energy in. After all, you have the do-not-disturb sign on your emotional cave 90 percent of the time. So, let's get real. You often need to hit full-blown crisis before you'll push yourself to get in the game and use your mighty transformative powers to grow past that pantheon of fears that are buried deep within you. Open up a little, let down your guard and cancel your security system. It's not really protecting you after all. It's just keeping you from discovering your deeper desires and living your passion. Trust that infamous Scorpionic radar to tell you what's really going on, and instead of deciding not to play anymore, change the rules of the game. Get rid of the crap; give yourself permission to live life wholeheartedly. For Scorpio, *the key to using change to create growth is purification.*

Sagittarius

You think your middle name is change, largely because you hate to be pinned down. You always keep all your options open and that leads you to believe that you love change. But, you carry around one hefty condition that keeps you from experiencing full-on enthusiasm that you are so famous for: Change is good as long as it doesn't lead to any kind of burden. That leaves you prone to passing up things that will truly benefit you because you mistake substance for inhibition. So all those lovely little adventures you are so fond of don't add the meaning to your life that you seek, because meaning only comes from investing something of yourself in the things you value. And no, this isn't a trick question. Or answer. It might seem like a double-edged sword, though. The challenge is to know that embracing what is meaningful to you truly does give you more freedom and that's what you need to grow. Forget running away because it ultimately leaves you with nothing. And that is your biggest fear—that life is about nothing at all. So, pick a destination. For Sagittarius, *the key to using change to create growth is discernment.*

Capricorn

If there's one thing in life that haunts you, it's change. You know what I am talking about. The kind that threatens the feeling that everything can be controlled. After all, if you don't believe that then, how can you achieve the goals you've set for yourself? Not for you the zigs or zags of life. Just straight lines please—the ones that take you from point A to point B with no diversions in between. What you fail to comprehend is that going off the beaten path can actually create a better end result than sticking to the script. However, that means you'd have to see change as something other than random, and that's hard for you. Try taking a deep breath. Step away from the "Über plan." Take your absolutely fabulous gift for sizing things up and plotting the right strategy, figure out what the unexpected is trying to show you and incorporate that into your overall approach. You'll grow more appreciation for your own skills and make good things happen. That's bound to be a whole lot more productive than moaning about the upside-down apple cart. For Capricorn, *the key to using change to create growth is flexibility.*

Aquarius

You believe change is necessary. In fact, you think you invented it. Heaven forbid that anything should ever be stagnant! But shaking things up for the sake of shaking them up doesn't always create change. Knocking down a wall isn't that productive unless you have something with which to replace it. Otherwise it's a lot like looking at Humpty Dumpty after his great fall. So let go of chaos and turmoil. Use what's been shoved in your face to get beyond the need to resist. Admit it. You resist anything that isn't your idea. And that includes change. No matter how much you embrace the idea, when change barges in you're still taken by surprise. You knew it was coming, but hey, not like this. You bluster and complain until you morph the whole thing into something you planned to do anyway. Well, make life easier. Welcome things without resistance. Ignore the messenger and pay attention to the message instead. Use your vaunted objectivity and your capacity to see outside the box to grow things, not just tear them apart. That way, people might actually listen to what you have to say. For Aquarius, *the key to using change to create growth is acceptance.*

Pisces

You are definitely the most fluid person on the planet. You change realities with ease and grace, often moment to moment. However, what you don't see is that dabbling in many dimensions can leave you out of the loop in this earthly one. And on Earth is where you live. So when the call for change comes, don't fly off to another part of the universe. Life ultimately may be transitory, but you still have to live it. And that means being fully engaged, not dropping in and out like a flower child. To grow, you need to be present. All that visionary capacity that you have will go to waste if you don't take time to focus. A dream is just a fantasy unless you do something about it. Put your considerable visualizing capacity to work. Use your magic to create growth. Sit down with your feet firmly planted on the ground and create the necessary framework to get the job done. After all, you can see it, can't you? For Pisces, *the key to using change to create growth is grounding.*

About Our Contributors

Alice DeVille is an internationally known astrologer, writer, and consultant. In her busy northern Virginia practice, she specializes in relationships, government affairs, career and change management, feng shui, real estate, and business management. She has worked under contract with the federal government as an executive coach and management consultant. Contact her at DeVilleAA@aol.com/

Lesley Francis is a professional astrologer, writer, journalist, and teacher, whose five planets in Aquarius have led her on a fascinating journey through life. Her study of astrology began thirty-four years ago, and it proved to be a passion and an avocation until ten years ago, when her journalistic career gave way to her new life as a professional astrologer, psychic, and teacher. She can be contacted at lesley_francis@hotmail.com/

Robin Ivy is a radio personality, educator, and astrologer in Portland, Maine. She fuses her passion for music and the metaphysical in "Robin's Zodiac Zone," a feature on her morning show on 94 WCYY, Portland's new rock alternative. Visit Robin's Web site at http://www.robinszodiaczone.com/

Shakti Navran is a professional jeweler with a lifelong interest in the metaphysical. She lives and works in Hawaii, and has been crafting personalized soul jewelry since 1977. Visit Shakti's Web site at http://www.shaktijewelry.com/

Bruce Scofield, a professional astrologer with an international clientele, specializes in both psychological and electional astrology. He is the author of fifteen books and is on the faculty of Kepler College. Visit his Web page at http://www.onereed.com/

Kaye Shinker lives in southern Louisiana. Kaye has earned the NCGR Level IV Certification in Counseling and Education as well as the ISAR Certification. She currently serves on the NCGR board of Examiners. Kaye loves to talk about everyone's favorite subject—money—the marketplace, where to look for bargains and what to avoid. Kaye's book, *The Textbook for Financial Astrology* can be purchased at http://www.astrologicalinvesting.com/